Charles Stedman Newhall

The Leaf-Collector's Handbook and Herbarium

An Aid in the Preservation and in the Classification of Specimen Leaves of the Trees

of Northeastern America

Charles Stedman Newhall

The Leaf-Collector's Handbook and Herbarium
An Aid in the Preservation and in the Classification of Specimen Leaves of the Trees of Northeastern America

ISBN/EAN: 9783337178024

Printed in Europe, USA, Canada, Australia, Japan

Cover: Foto ©berggeist007 / pixelio.de

More available books at **www.hansebooks.com**

THE LEAF-COLLECTOR'S

HAND-BOOK and HERBARIUM

AN AID IN THE PRESERVATION AND IN THE CLAS-
SIFICATION OF SPECIMEN LEAVES OF THE
TREES OF NORTHEASTERN AMERICA

BY

CHARLES S. NEWHALL

AUTHOR OF "THE TREES OF NORTHEASTERN AMERICA," ETC.

ILLUSTRATED

G. P. PUTNAM'S SONS

NEW YORK   LONDON
27 & 29 WEST TWENTY-THIRD ST.   27 KING WILLIAM ST., STRAND

The Knickerbocker Press

1891

CONTENTS.

ILLUSTRATIONS.

PREFACE.

The leaves described and pictured in the following pages represent all the native trees, and the most important introduced and naturalized trees of Northeastern America,*—with the few exceptions named on page 213.

Those species are considered trees (in distinction from shrubs) which, as a rule, spring from the ground with a single branching trunk.†

How to use the book will be readily understood by consulting the following directions.

* The names of introduced and naturalized species are enclosed in brackets.
† "Trees of Northeastern America," page 4.

DIRECTIONS.

1. *How to find the names of specimens.*—Compare any given specimen, first with the descriptions in the "Guide," on pages xiv and xv, and then with the illustrations to which the "Guide" directs.*

2. *How to mount specimens.*—When the specimens have been pressed and *thoroughly dried*, and all their thick stems removed or pared, they should be fastened in their places opposite the corresponding illustrations, with strips of gummed paper an eighth of an inch or less in width.†

The compound leaves and the simple leaves, when they are larger than the allotted space, should be represented by sections of the leaf.

3. *How to preserve specimens.*—If the collection is attacked by insects, each leaf should be brushed lightly with a *saturated solution of corrosive sublimate and alcohol, increased by two thirds more of alcohol.*

4. *Notes.*—Interest and value will be added to the collection if full memoranda are kept of dates, localities, name of the finder, incidents, characteristics of the tree, etc.

5. It should be remembered that leaves from vigorous young sprouts are not usually the best specimens. It is seldom that two leaves, even upon the same mature branch, exactly agree, but they follow the type, while often the younger growth varies from it.‡

* If fuller descriptions and comments are desired, they can be found in the author's work on "The Trees of Northeastern America."
† A supply of gummed paper will be found at the end of the volume.
‡ "The Trees of Northeastern America," page 4.

xi

LIST OF GENERA.

GUIDE.

Leaves, simple :*
- **alternate†**
  - edge entire. Go to A—*I*
  - " toothed. " A—*II*
  - " lobed
    - Lobes, entire. Go to A—*III (a)*
    - Lobes, toothed. " A—*III (b)*
- **opposite**
  - edge entire. Go to B—*I*
  - " toothed. " B—*II*
  - " lobed
    - Lobes, entire. Go to B—*III (a)*
    - Lobes, toothed. " B—*III (b)*
- Indeterminate. Go to C—*I*

Leaves, compound :
- **feather-shaped ‡**
  - alternate, edge
    - entire. Go to D—*I*
    - toothed. " D—*II*
  - opposite, edge
    - entire. Go to E—*I*
    - toothed. " E—*II*
- **hand-shaped §**
  - opposite, edge toothed. Go to F—*I*

---

* The leaflets of a compound leaf can be distinguished from a simple leaf by the absence of leaf-buds from the base of their stems.

† Referring to the arrangement of the leaves on the branch.

‡ *E. g.*, as in the hickories, sumachs, etc.

§ *E. g.*, as in the horse-chestnut, etc.

GUIDE (*Continued*).

NOTE.—Names in *italics* are given also under another division.

TREES WITH SIMPLE LEAVES

LEAVES ALTERNATE

(EDGE ENTIRE)

A I

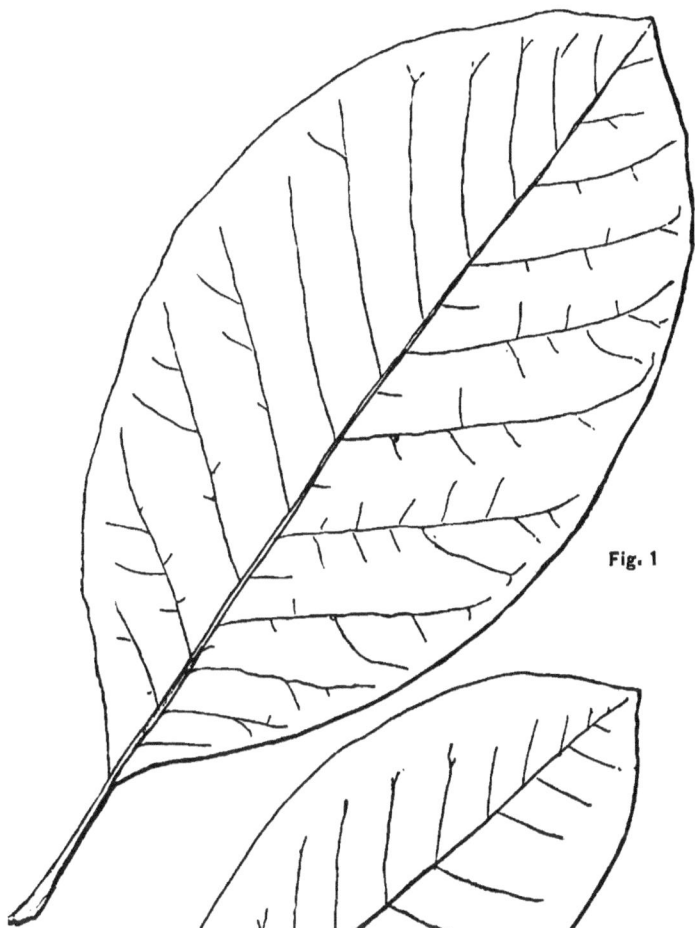

Fig. 1

Fig. 2

Fig. 1.—Cucumber Tree. (M. acumìnàta, L.)
Fig. 2.—Sweet Bay. (M. glauca, L.)

3   NATURAL SIZE.

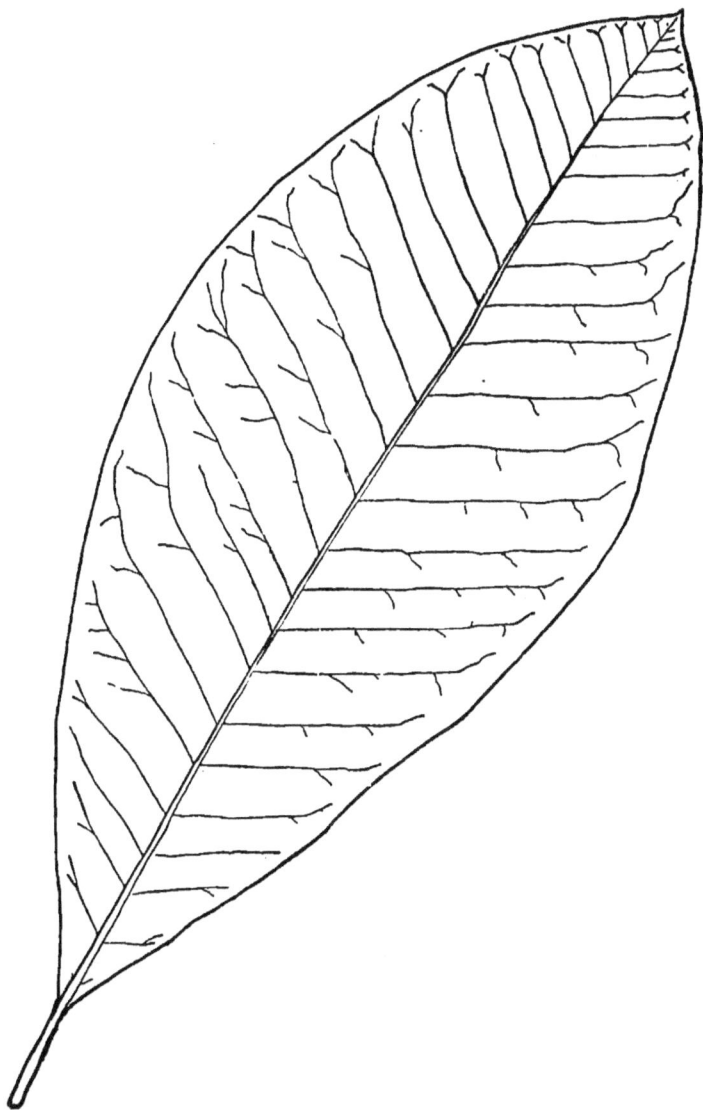

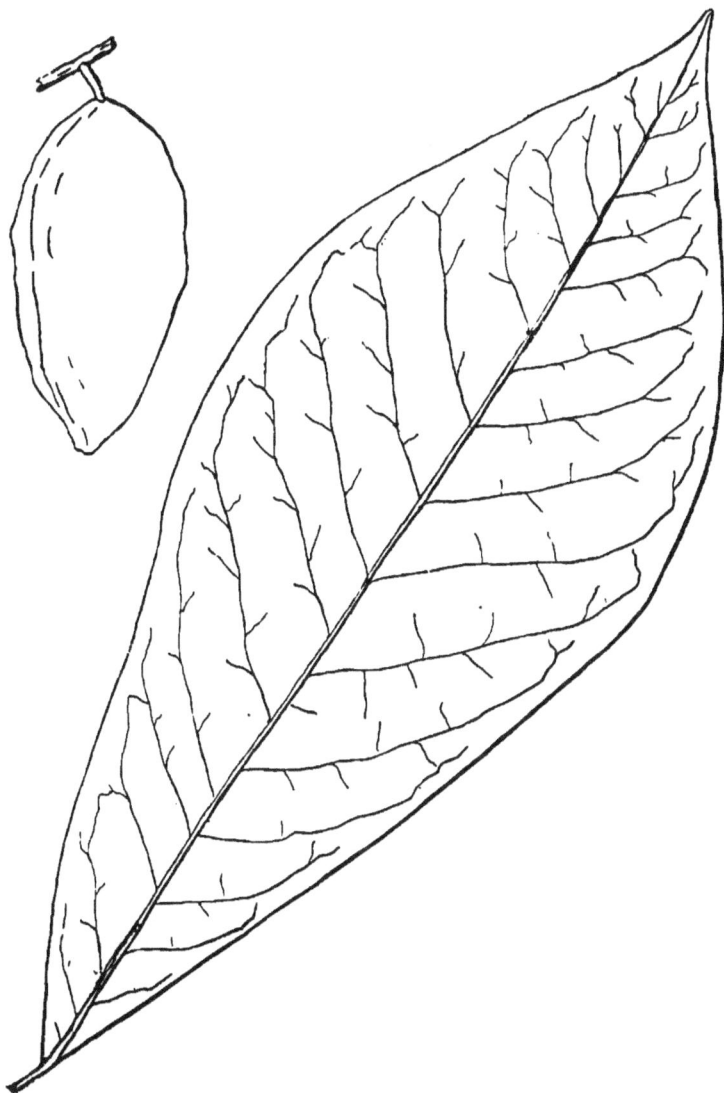

Fig. 4.—Papaw. A. triloba (L.), Dunal.
LEAF, NATURAL SIZE. FRUIT, TWO THIRDS NATURAL SIZE.

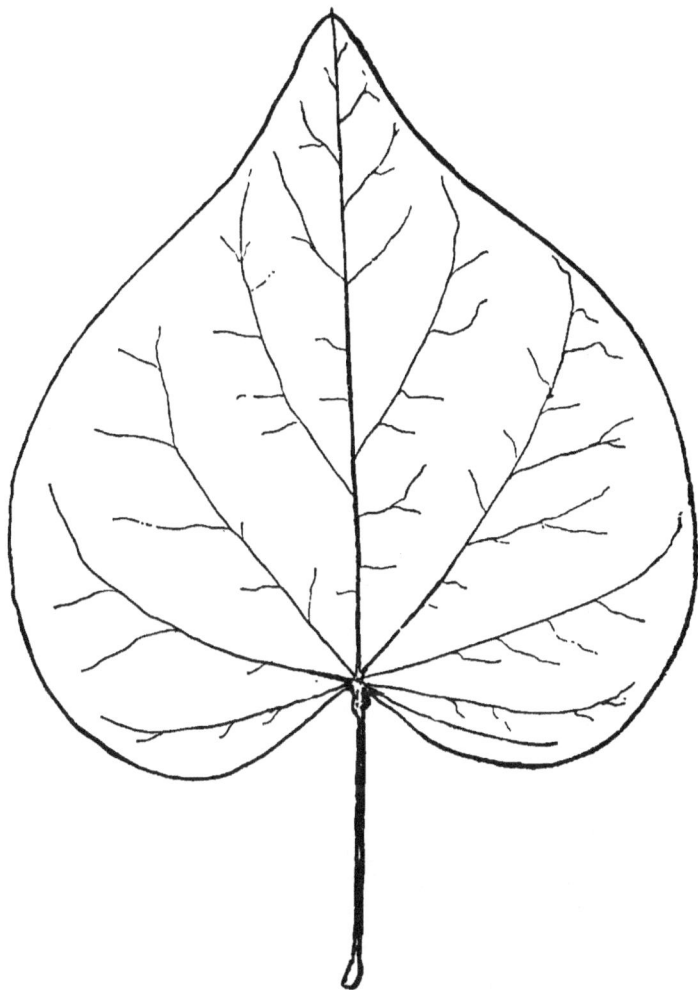

Fig. 5.—Red Bud. (C. Canadènsis, L.)
NATURAL SIZE.

Fig. 6.—Sour Gum.　(N. sylvàtica, Marsh.)
NATURAL SIZE.

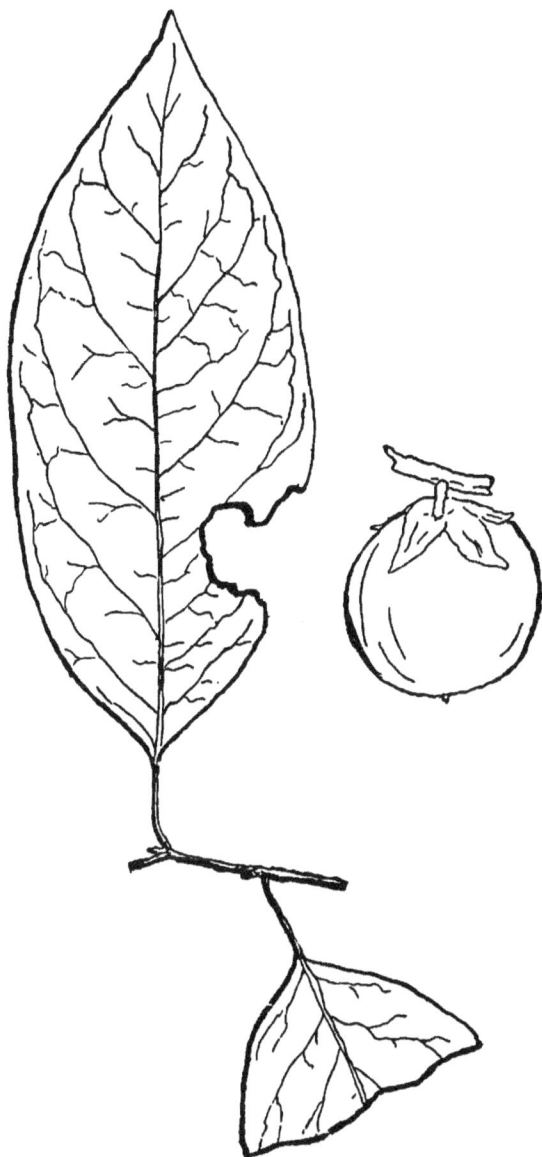

Fig. 7.—Persimmon. (D. Virginiàna, L.)
NATURAL SIZE.

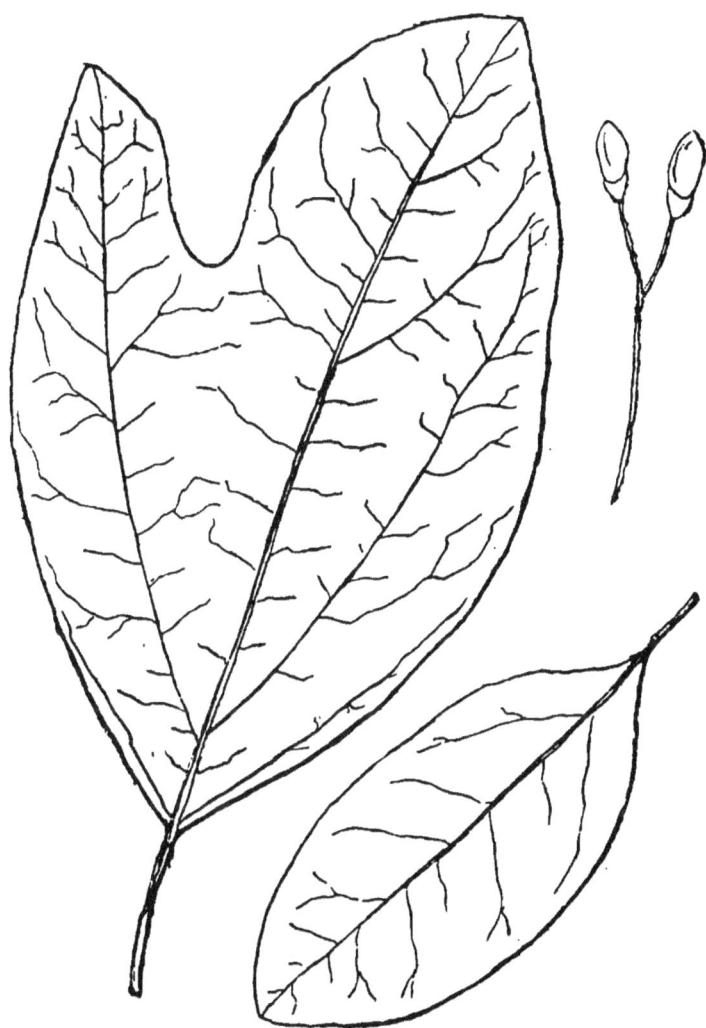

Fig. 8.—**Sassafras.** (S. officinàle, Nees.)
NATURAL SIZE.

TREES WITH SIMPLE LEAVES

LEAVES ALTERNATE
CONTINUED

(EDGE TOOTHED)

A II

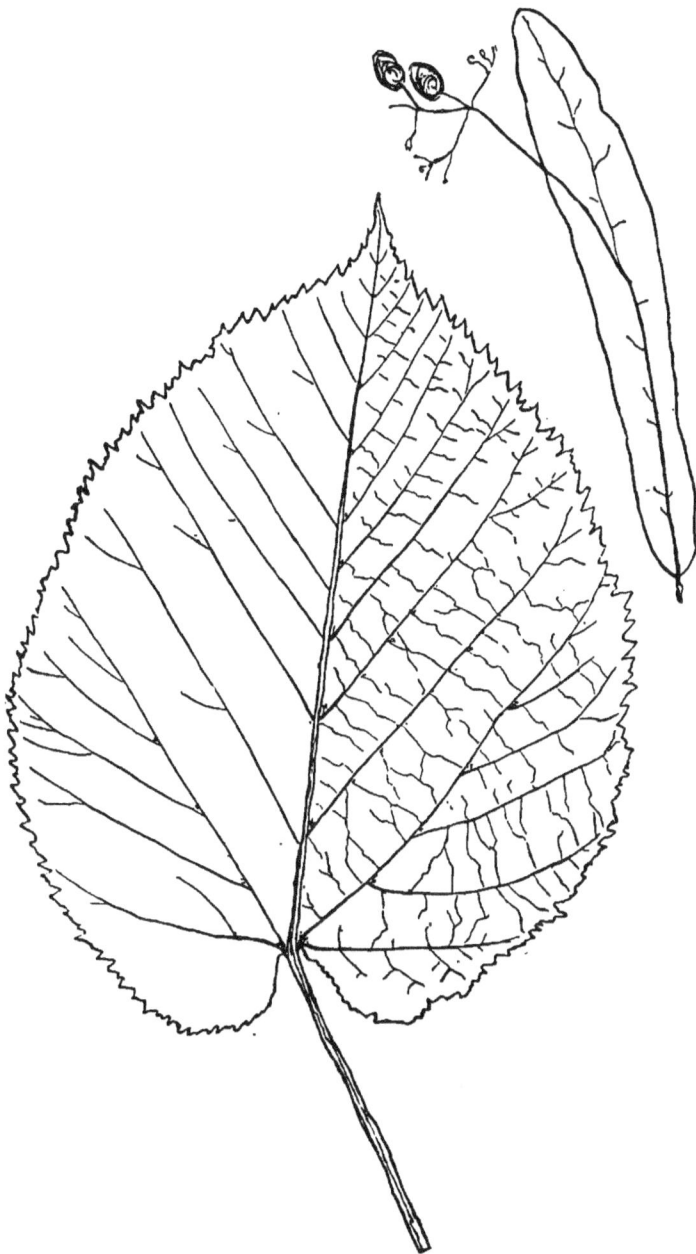

Fig. 9.—Basswood. (T. Americàna, L.)

NATURAL SIZE.

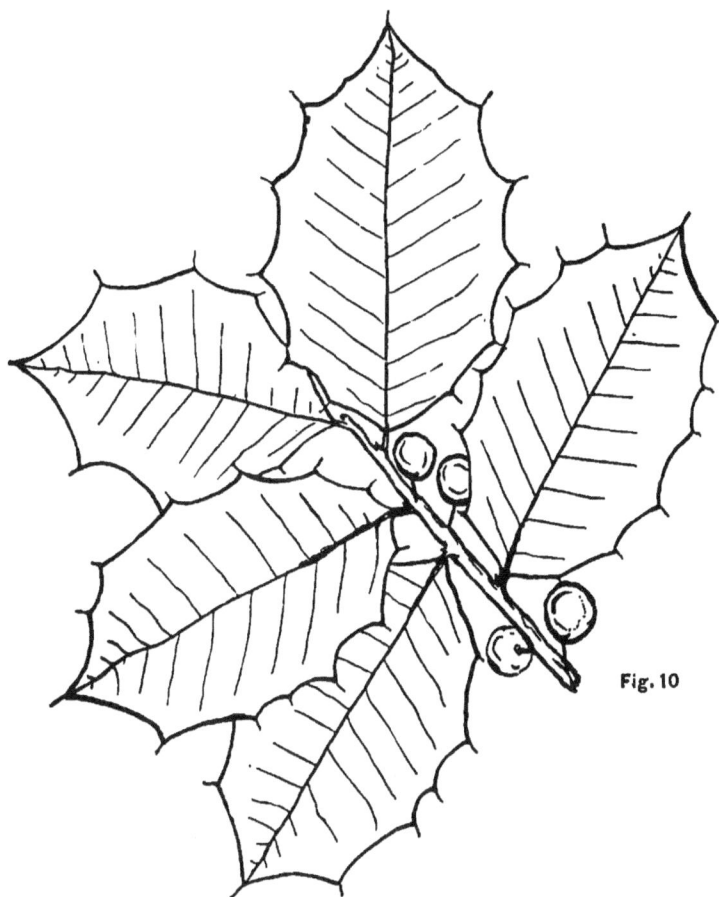

Fig. 10

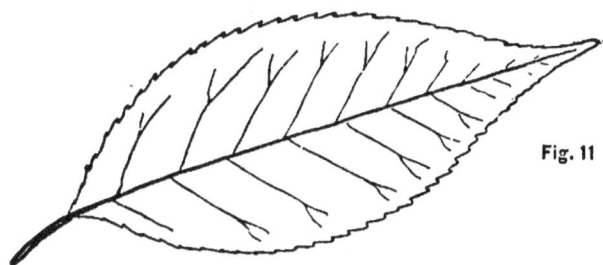

Fig. 11

Fig. 10.—American Holly. (I. opàca, Ait.)
Fig. 11.—I. montìcola.
NATURAL SIZE.

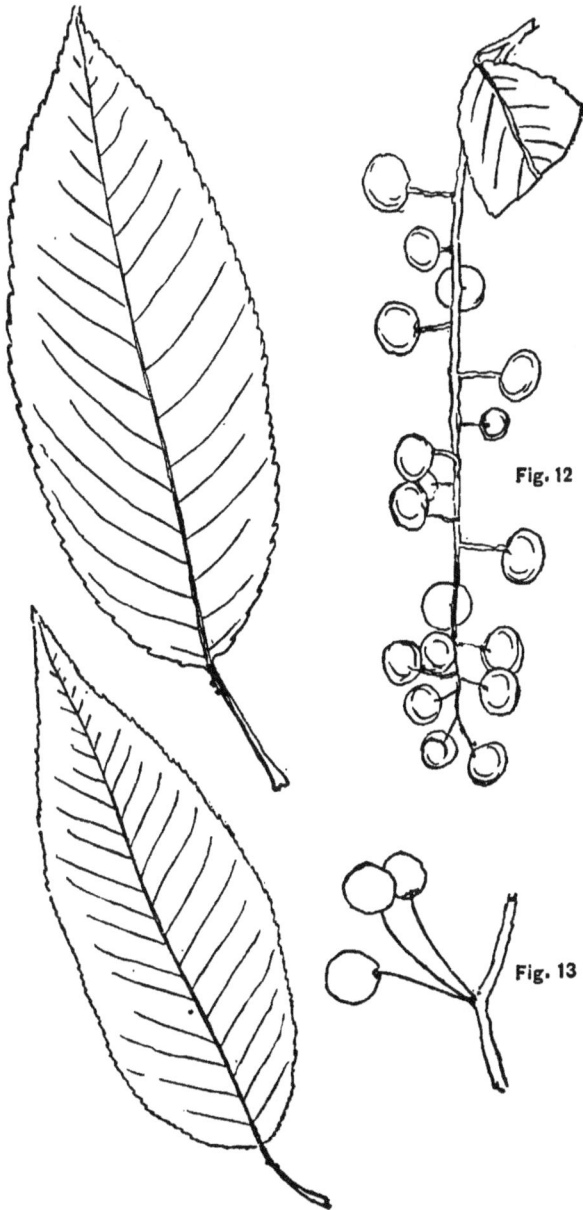

Fig. 12

Fig. 13

Fig. 12.—Wild Black Cherry.   (P. serótina, Ehr.)
Fig. 13.—Wild Red Cherry.   (P. Pennsylvànica, L).

NATURAL SIZE.

23

Fig. 14.—Wild Plum. (P. Americàna, Marsh.)
NATURAL SIZE.

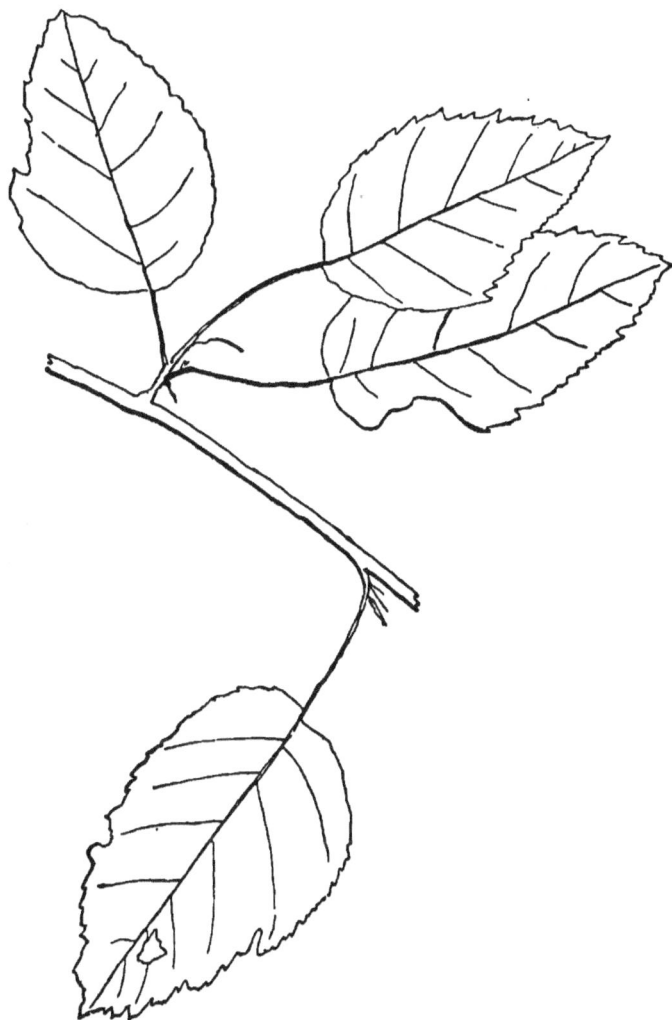

Fig. 15.—Crab-Apple. (P. coronària, L.)
NATURAL SIZE.

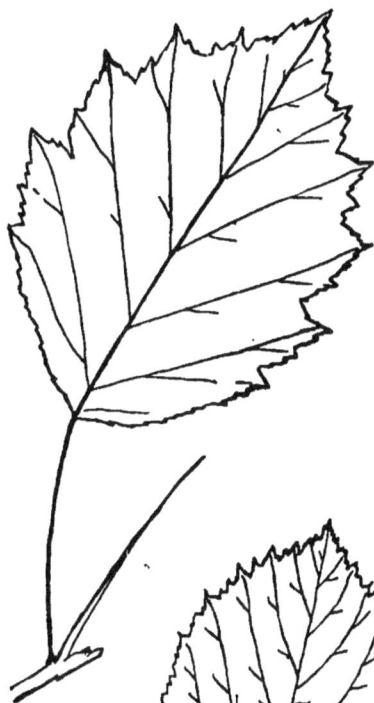

Fig. 16

Fig. 17

Fig. 16.—White Thorn and Fruit. (C. coccínea, L.)
Fig. 17.—Black Thorn. (C. tomentòsa, L.)
NATURAL SIZE.

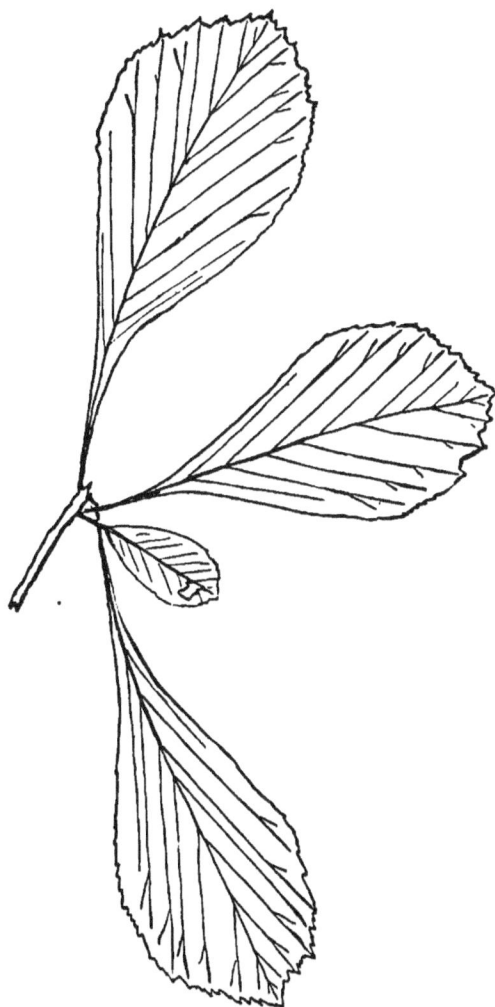

Fig. 18.—Common Thorn. (C. punctàta, Jac.)
NATURAL SIZE.

Fig. 19.—Cockspur Thorn. (C. crus-galli, L.)
NATURAL SIZE.

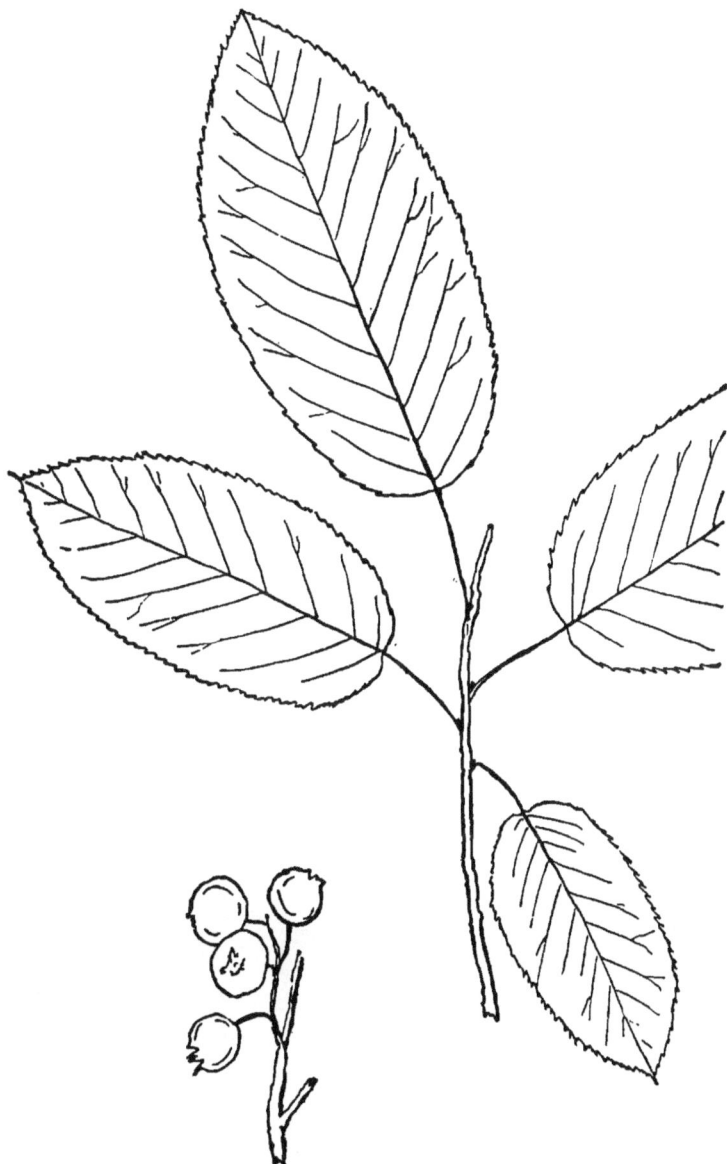

Fig. 20.—Shad-bush. A. Canadènsis (L.), Medik.
NATURAL SIZE.

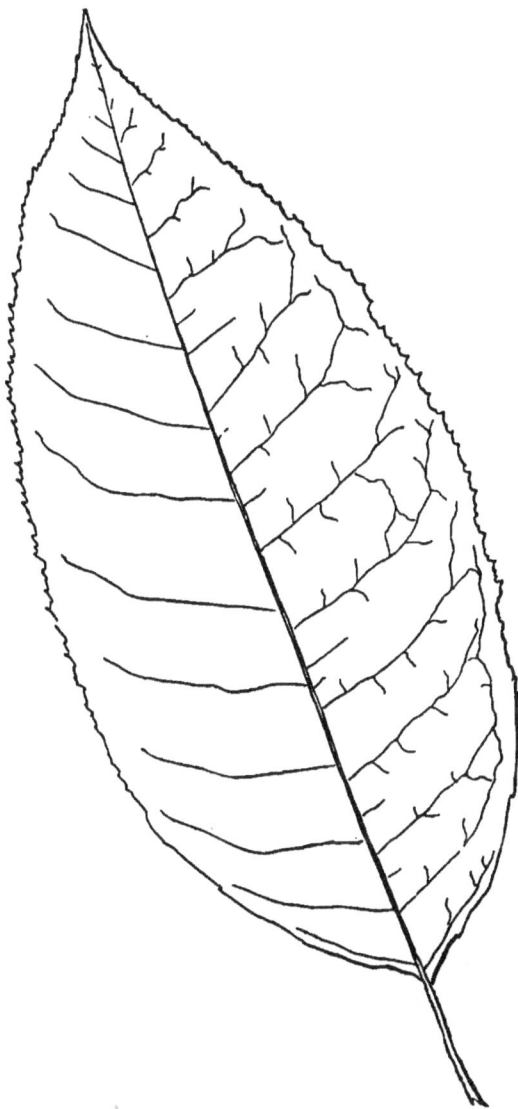

Fig. 21.—Sorrel Tree.   O. arbòreum (L.), D. C.
NATURAL SIZE.

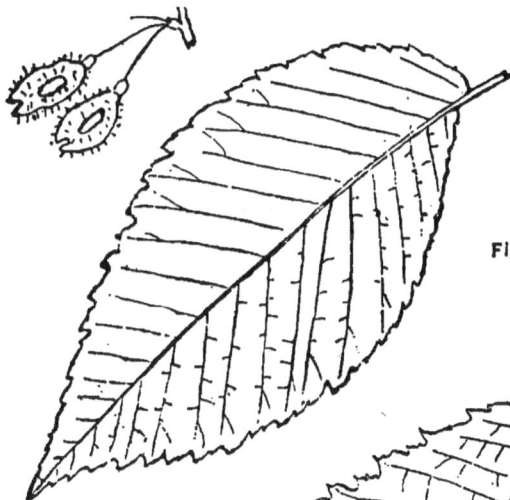

Fig. 22

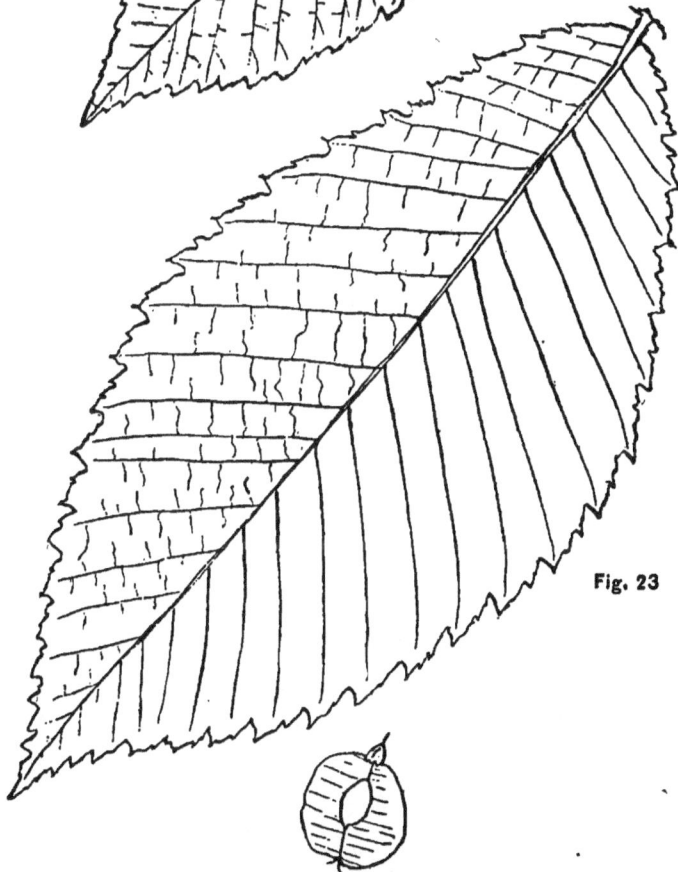

Fig. 23

Fig. 22.—White Elm. (U. Americàna, L.)
Fig. 23.—Slippery Elm. (U. fulva, Michaux.)
NATURAL SIZE.

Fig. 24.—Hackberry. (C. occidentàlis, L.)
NATURAL SIZE.

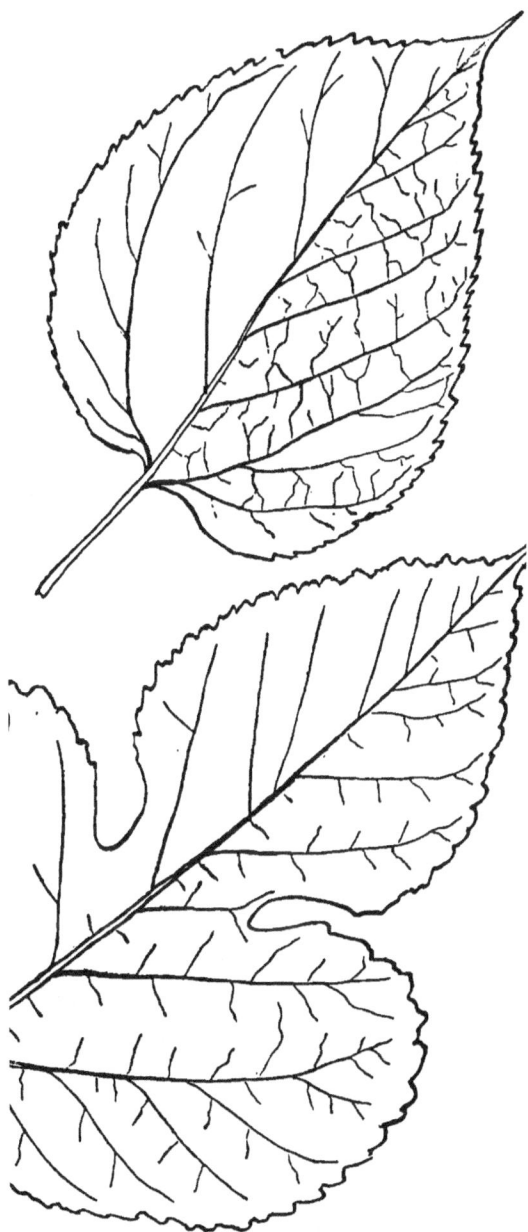

Fig. 25.—Red Mulberry. (M. rubra, L.)
NATURAL SIZE.

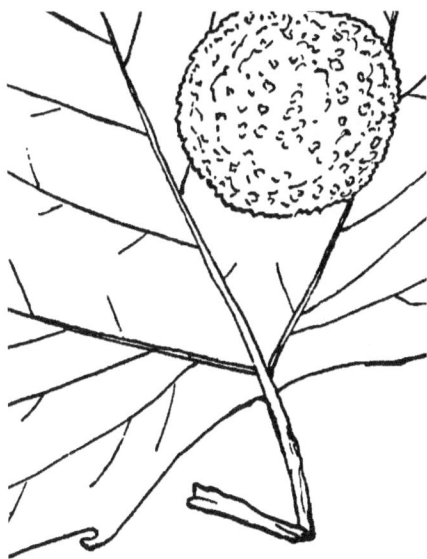

Fig. 26.—Buttonwood. (P. occidentàlis, L.)
NATURAL SIZE.

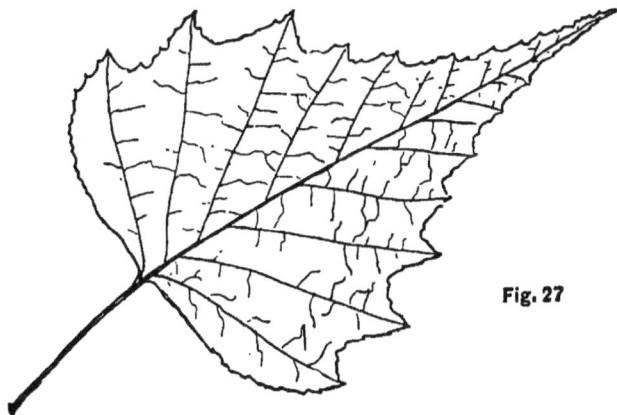

Fig. 27

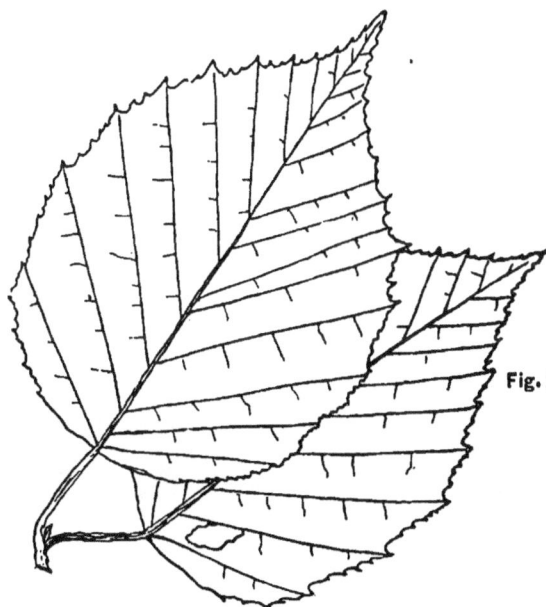

Fig. 28

Fig. 27.—White Birch. (B. populifòlia, Marsh.)
Fig. 28.—Paper Birch. (B. papyrìfera, Marsh.)
NATURAL SIZE.

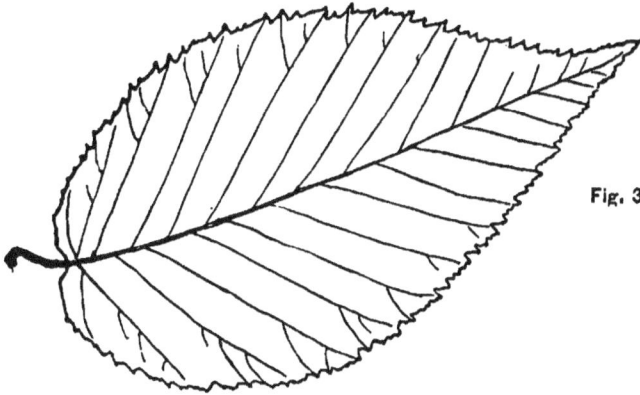

Fig. 29

Fig. 30

Fig. 29.—Red Birch. (B. nigra, L.)
Fig. 30.—Yellow Birch. (B. lùtea, Michaux, f.)
NATURAL SIZE.

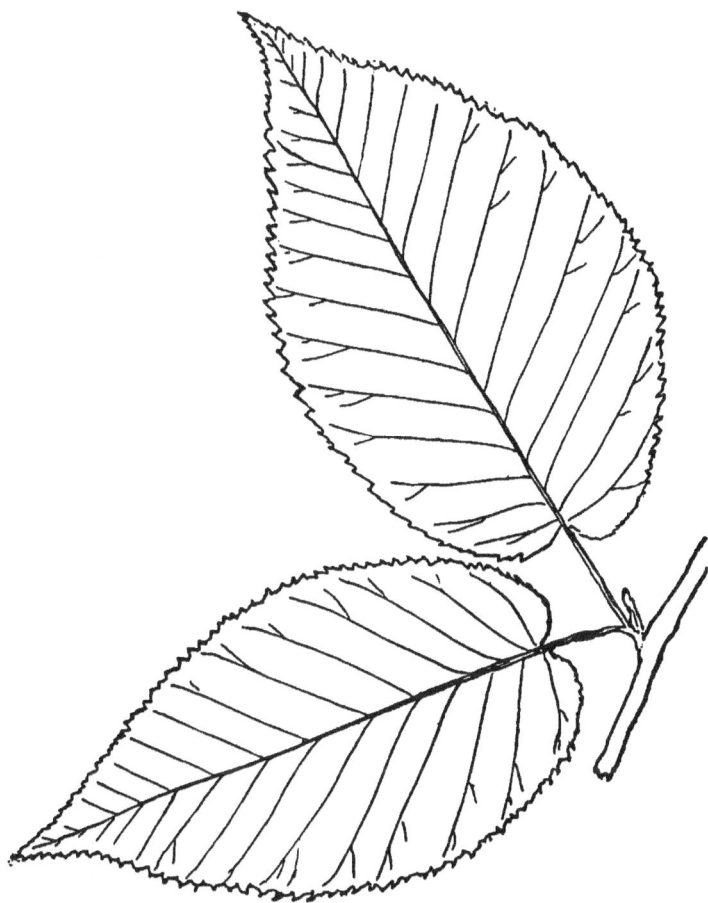

Fig. 31.—Sweet Birch. (B. lenta, L.)
NATURAL SIZE.

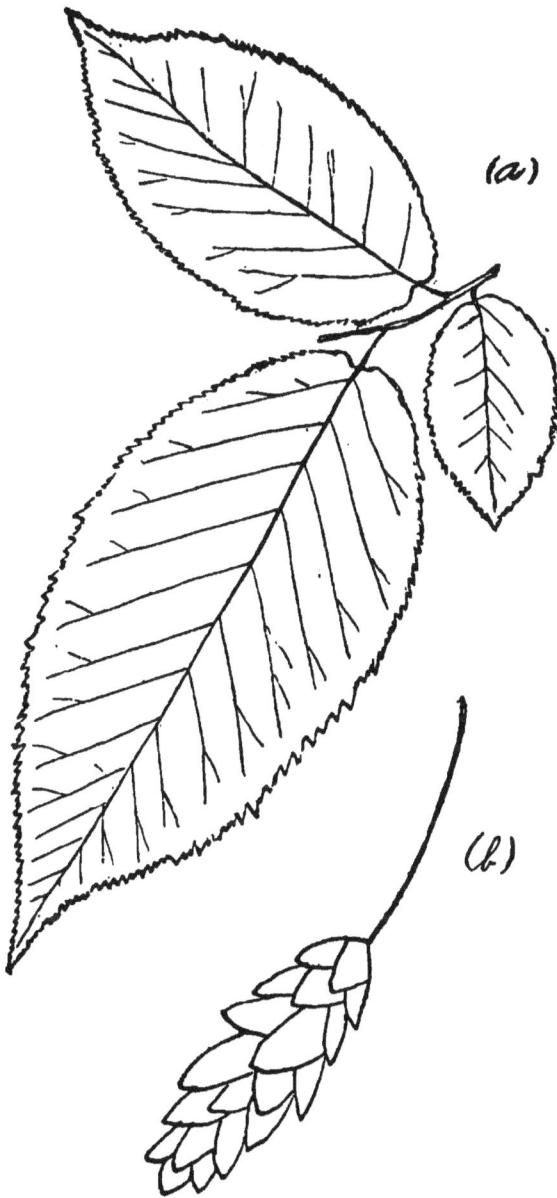

Fig. 32.—Hop-Hornbeam.   O. Virginiàna (Mill), Willd.
a. Leaves.   *b*. Fruit.

NATURAL SIZE.

(*a*)

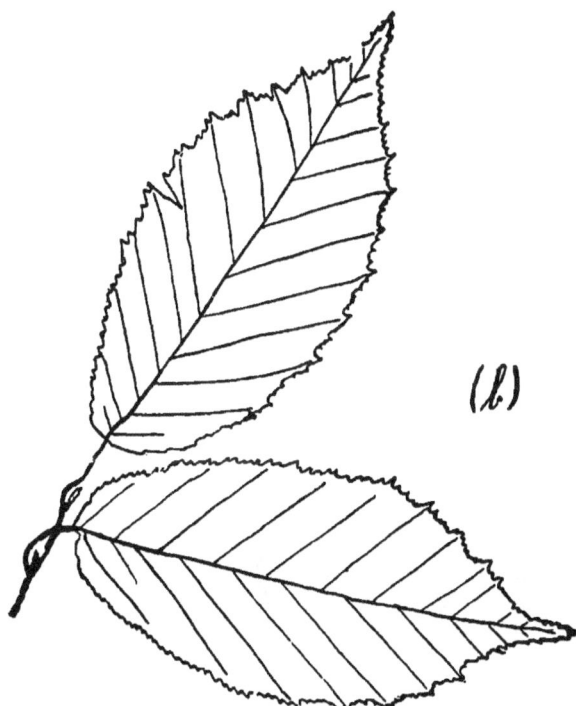

(*b*)

Fig. 33.—Hornbeam. (C. Caroliniàna, Walt.)

a. Fruit scales. *b*. Leaves.

NATURAL SIZE.

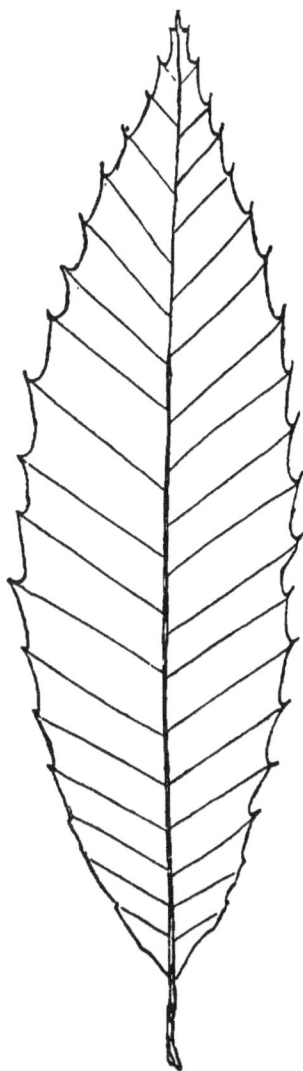

Fig. 34.—Chestnut. C. Satlva (L.), var. Americàna (Michaux), Sarg.
NATURAL SIZE.

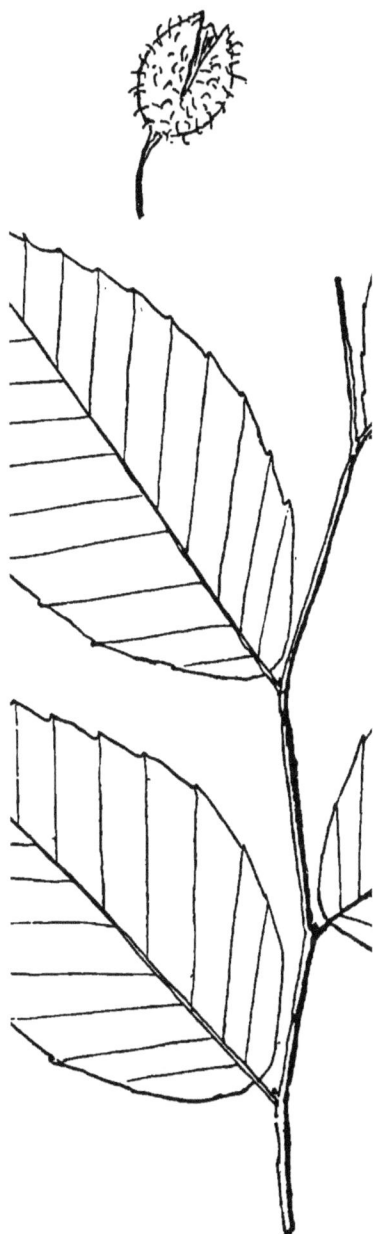

Fig. 35.—Beech. (F. ferruginea, Ait.)
NATURAL SIZE.

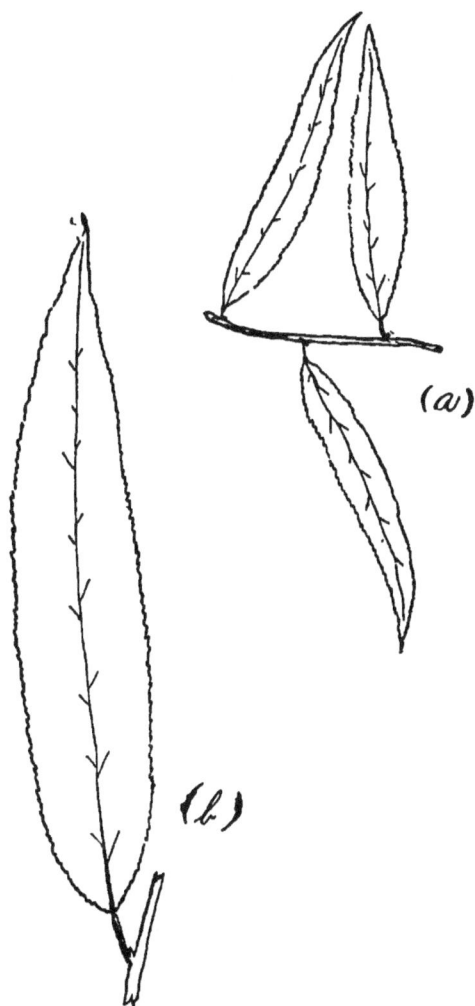

Fig. 36.—Black Willow. (S. nigra, Marsh.)
a. Commonest form. *b.* Large form.
NATURAL SIZE.

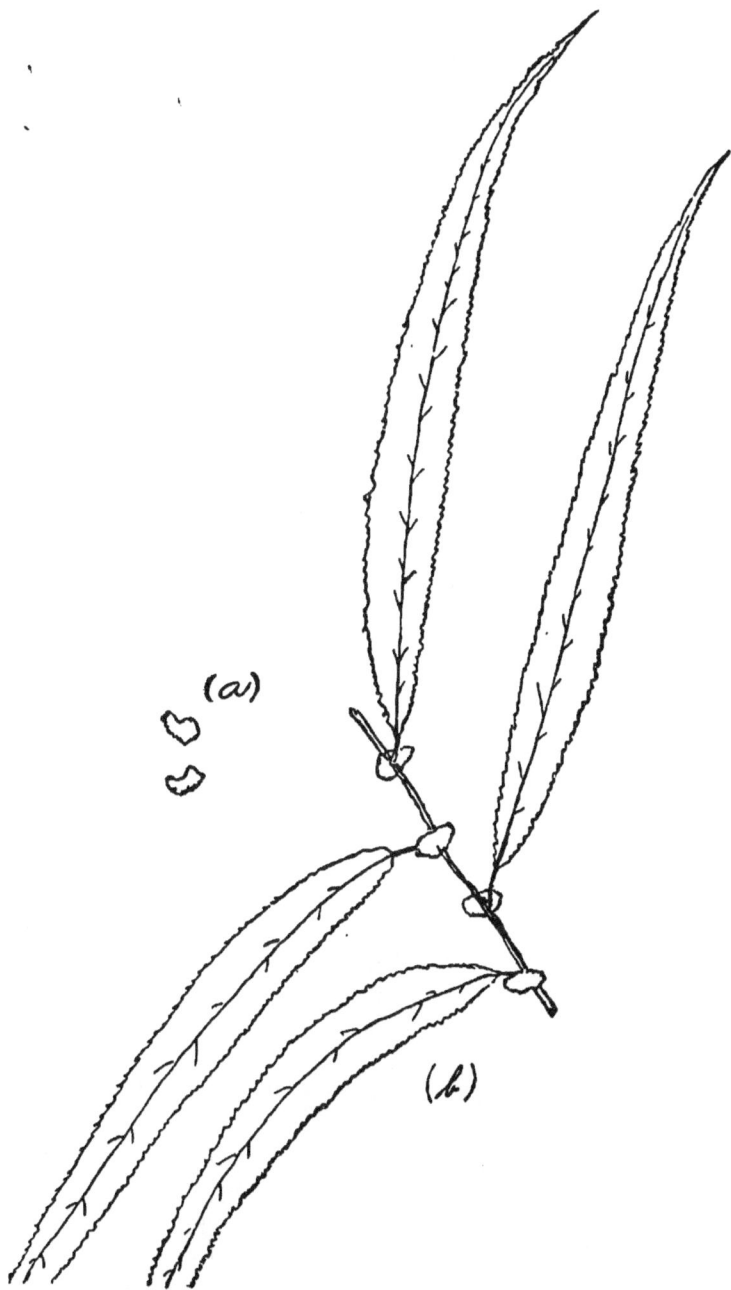

(a)

(b)

Fig. 37.—Scythe-leaved Willow. (S. n., var. falcàta, Torr.)

a. Stipules. b. Leaves.

63

NATURAL SIZE.

Fig. 38.—Shining Willow.　(S. lùcida, Muhl.)

NATURAL SIZE.

''

Fig. 39.—Long-beaked Willow. (S. rostràta, Richards.)
NATURAL SIZE.

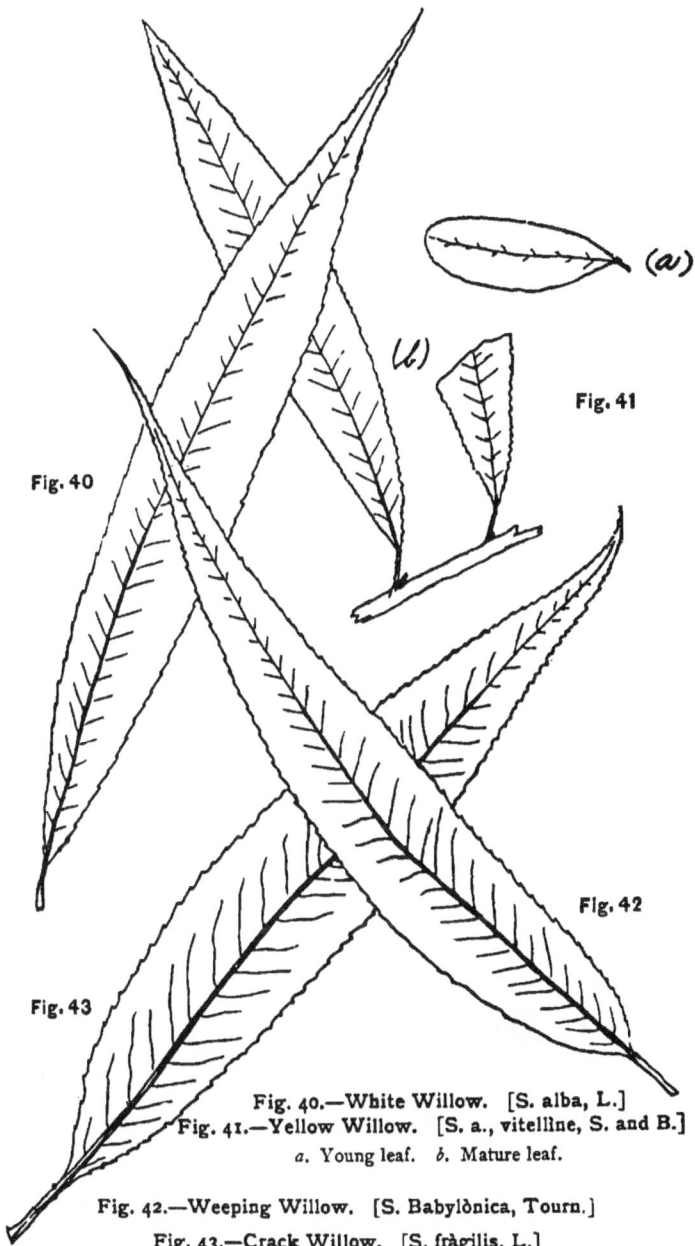

Fig. 40.—White Willow. [S. alba, L.]
Fig. 41.—Yellow Willow. [S. a., vitelline, S. and B.]
a. Young leaf. b. Mature leaf.

Fig. 42.—Weeping Willow. [S. Babylònica, Tourn.]
Fig. 43.—Crack Willow. [S. fràgilis, L.]
NATURAL SIZE.

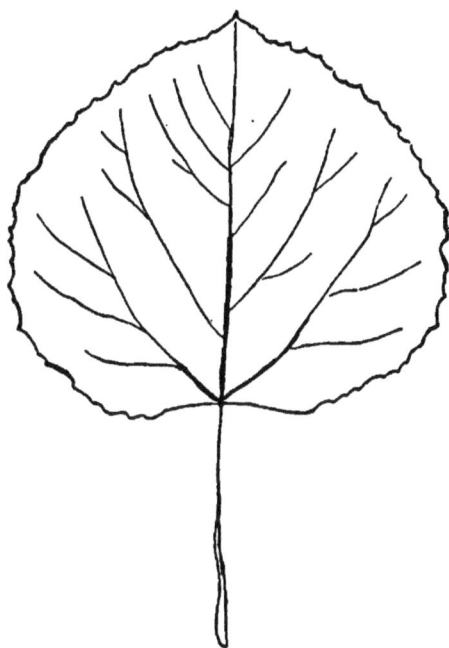

Fig. 44.—Aspen. (P. tremuloides, Michx.)
NATURAL SIZE.

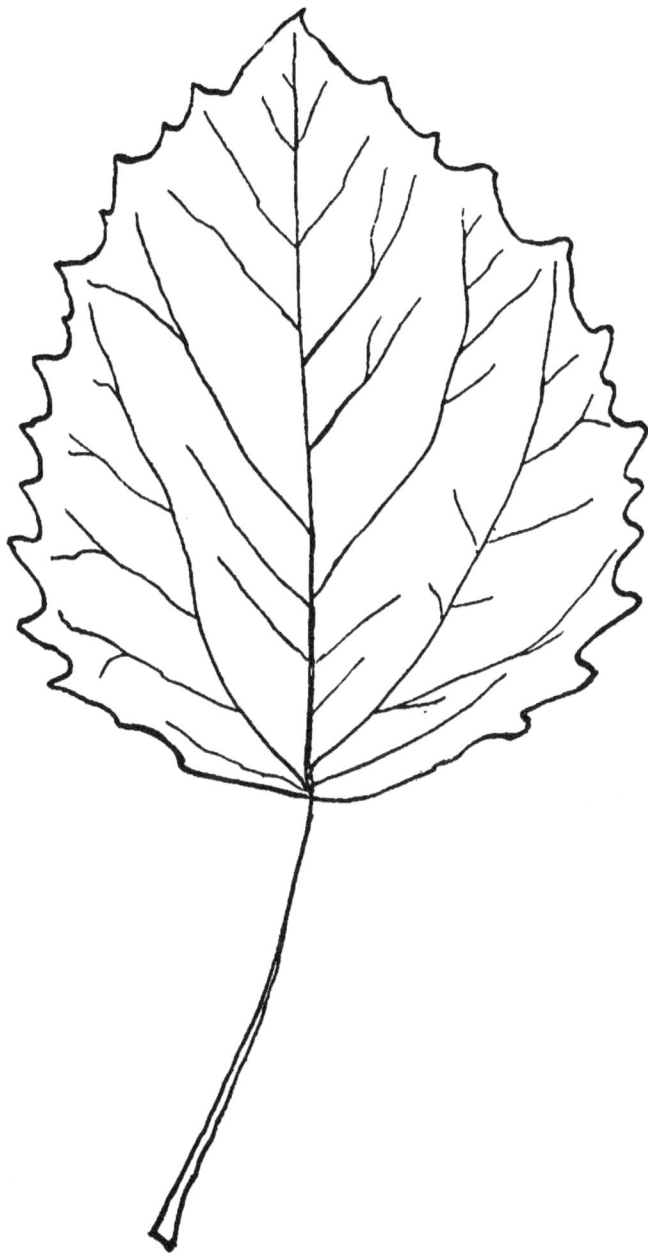

Fig. 45.—Large Toothed Aspen. (P. grandidentàta, Michx.)

NATURAL SIZE.

Fig. 46.—Downy-leaved Poplar. (P. heterophýlla, L.)
NATURAL SIZE.

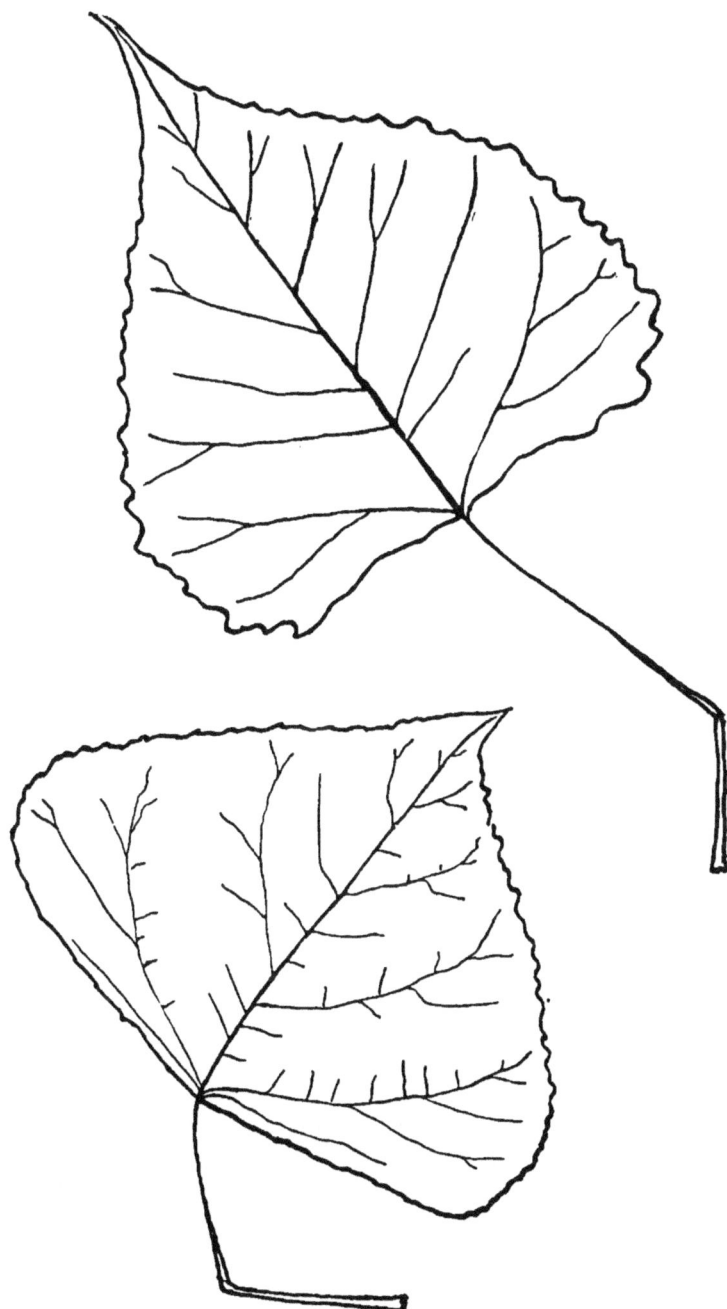

Fig. 47.—Cottonwood. (P. monilifera, Ait.)
NATURAL SIZE.

Fig. 48.—Balsam Poplar.   (P. balsamifera, L.)
Fig. 49.—Balm of Gilead.   P. b. càndicans (Ait.), Gray.

NATURAL SIZE.

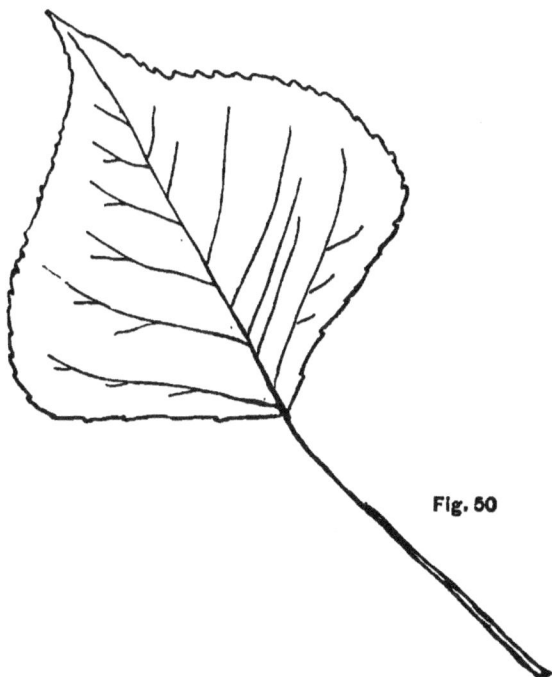

Fig. 50

Fig. 51

Fig. 50.—Lombardy Poplar. [P. dilatàta, Ait.]
Fig. 51.—Silver-Leaf Poplar. [P. alba, L.]

NATURAL SIZE.

TREES WITH SIMPLE LEAVES

LEAVES ALTERNATE

CONTINUED

(EDGE LOBED)

A III

(a) and *(b)*

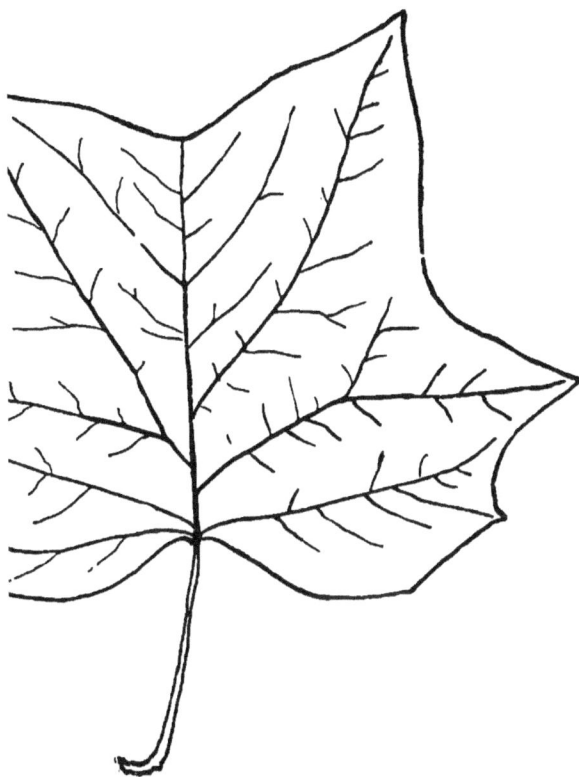

Fig. 52.—Tulip Tree. (L., tulipifera, L.)
NATURAL SIZE.

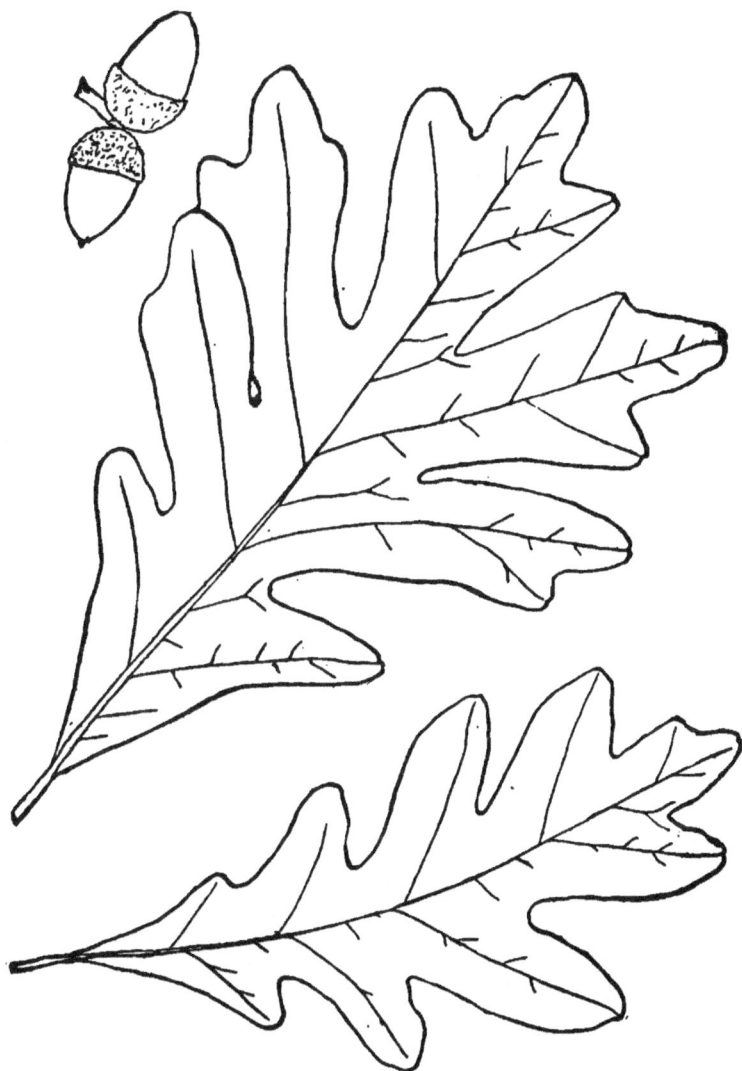

Fig. 53.—White Oak. (Q. alba, L.)
LEAVES AND FRUIT REDUCED ONE FOURTH.

Fig. 54.—Post Oak. Q. minor (Marsh), Sarg.
LEAVES AND FRUIT REDUCED ONE FOURTH.

Fig. 55.—Burr Oak. (Q. macrocàrpa, Michx.)
LEAVES AND FRUIT REDUCED ONE FOURTH.

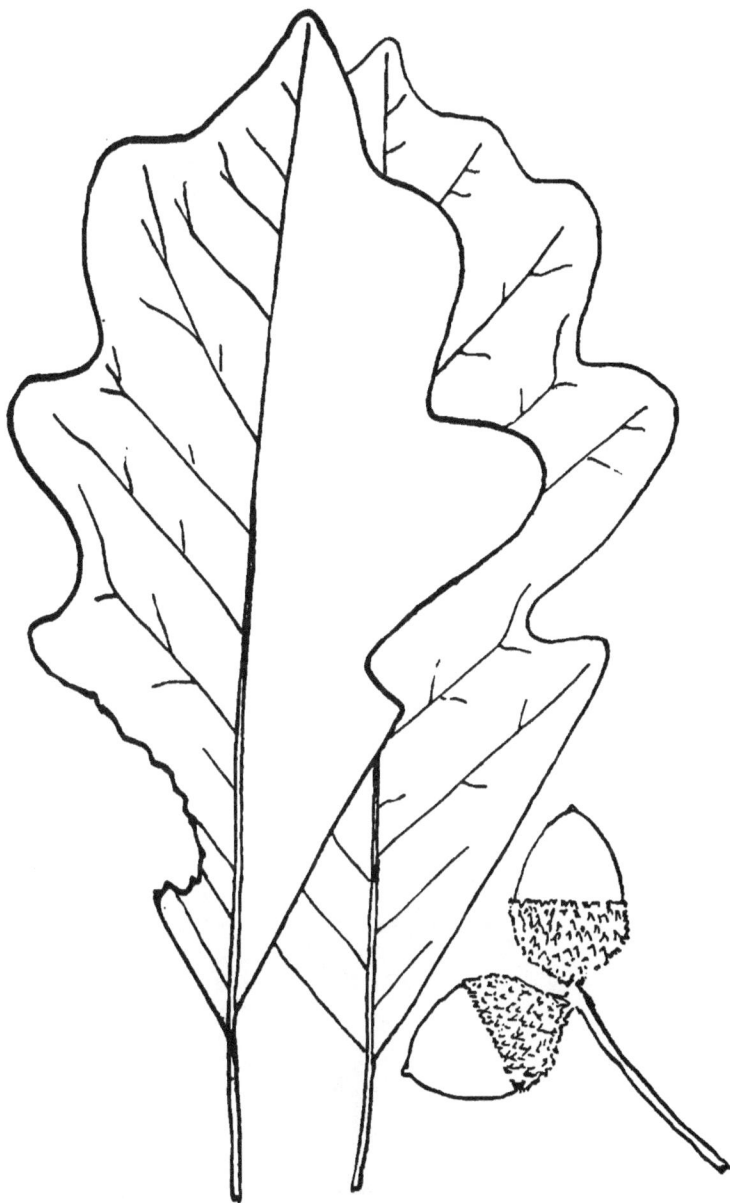

Fig. 56.—Swamp White Oak.   (Q. bicolor, Willd.)
NATURAL SIZE.

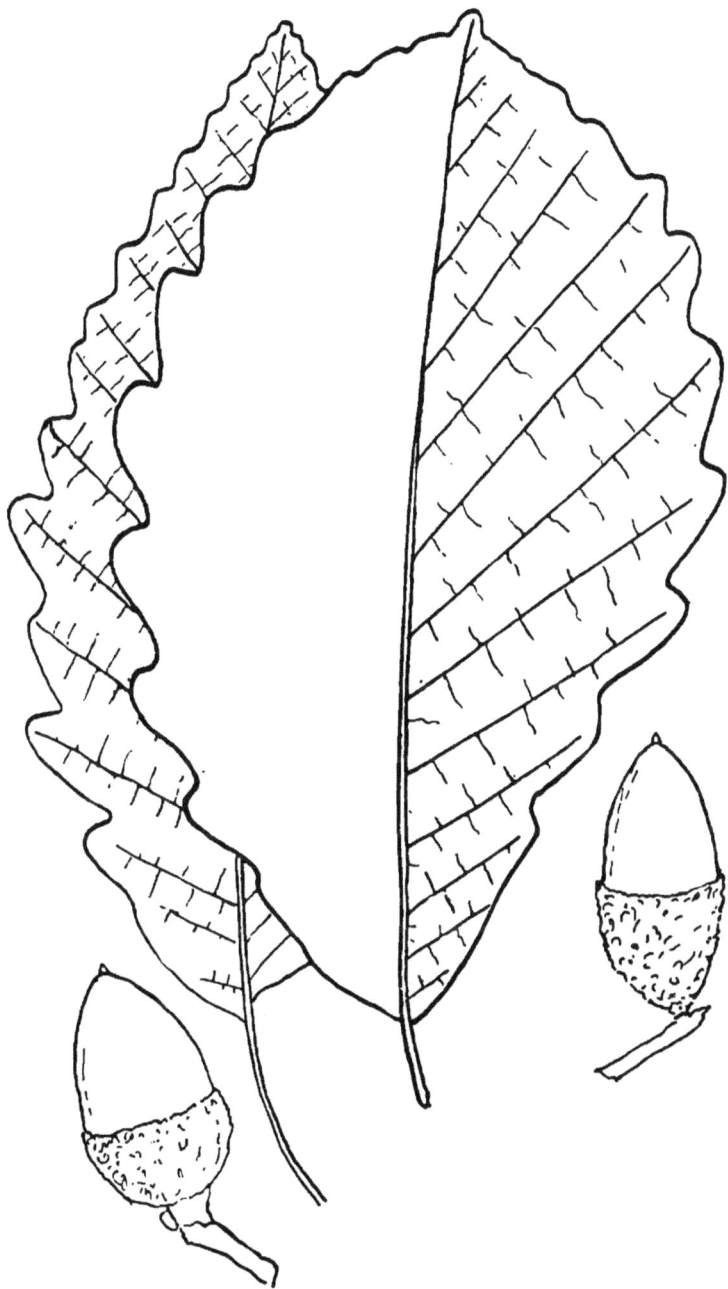

Fig. 57.—Chestnut Oak. (Q. prinus, L.)
NATURAL SIZE.

Fig. 58.—Yellow Chestnut Oak.   Q. (Muhl.), Engel.
NATURAL SIZE.

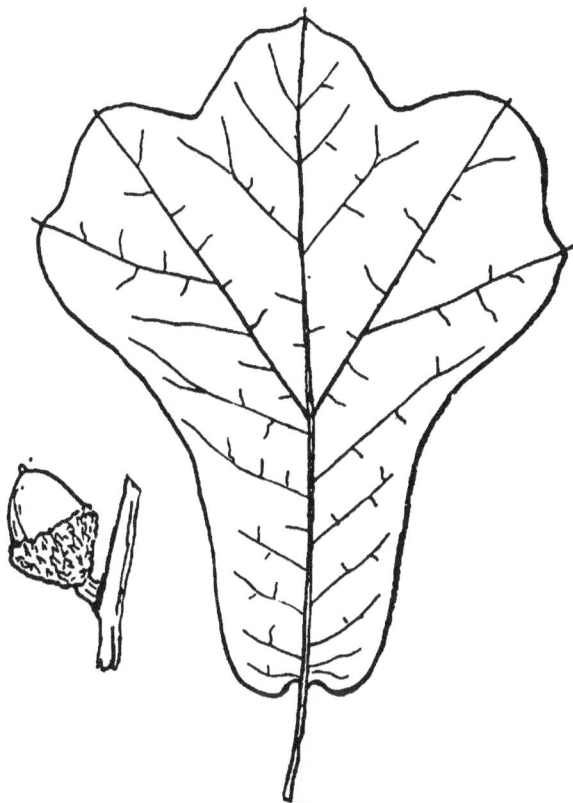

Fig. 59.—Black Jack. (Q. nigra, L.)
NATURAL SIZE.

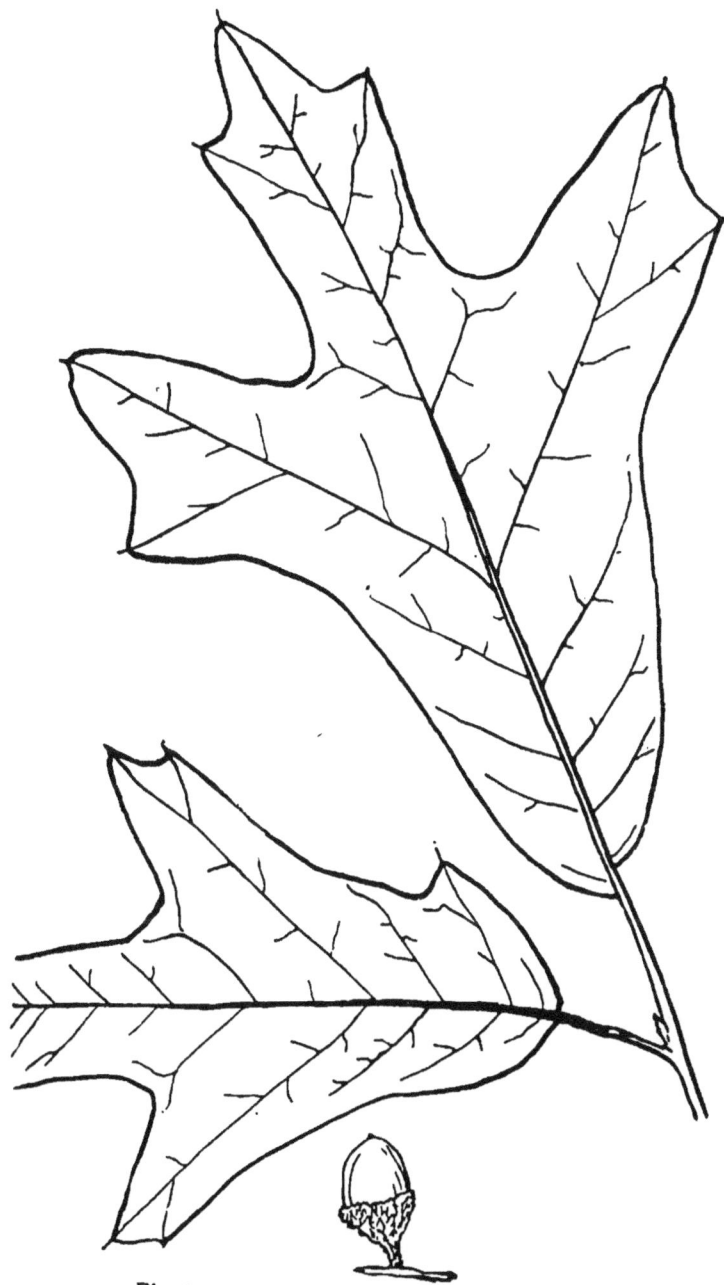

Fig. 60.—Spanish Oak. (Q. cuneàta, Wang.)
NATURAL SIZE.

Fig. 61.—Scarlet Oak. (Q. coccinea, Wang.)
NATURAL SIZE.

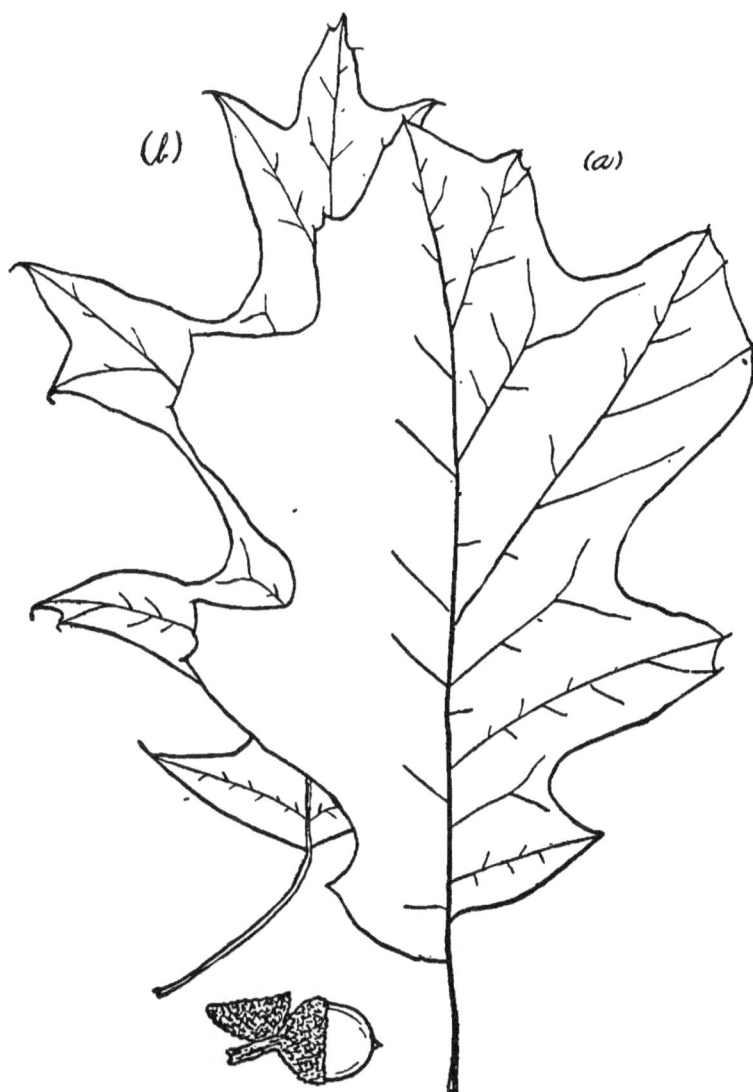

(b) *(a)*

Fig. 62, *a* and *b*.—**Black Oak.** (Q. c., tinctòria, Gray.)
FRUIT AND LEAVES REDUCED ONE FOURTH.

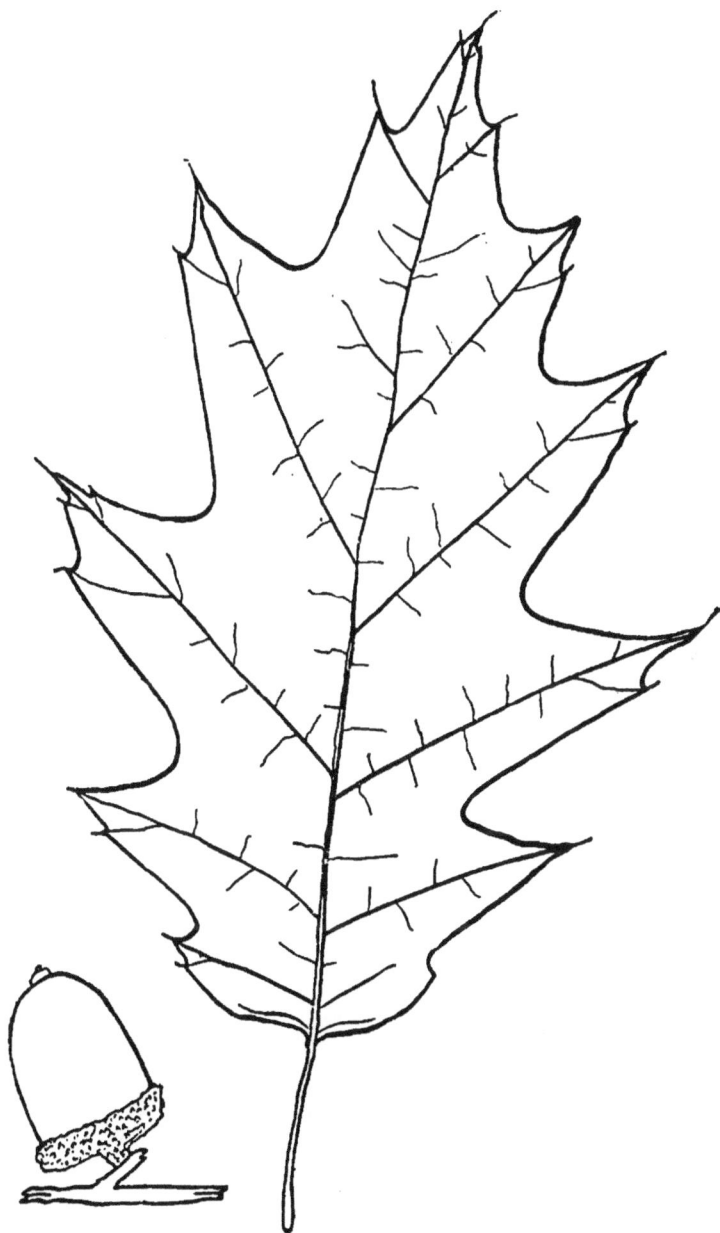

Fig. 63.—Red Oak. (Q. rubra, L.)
NATURAL SIZE.

Fig. 64.—Pin Oak. (Q. palùstris, D. Roi.)
NATURAL SIZE.

Fig. 65.—Willow Oak. (Q. Phellos, L.)
NATURAL SIZE.

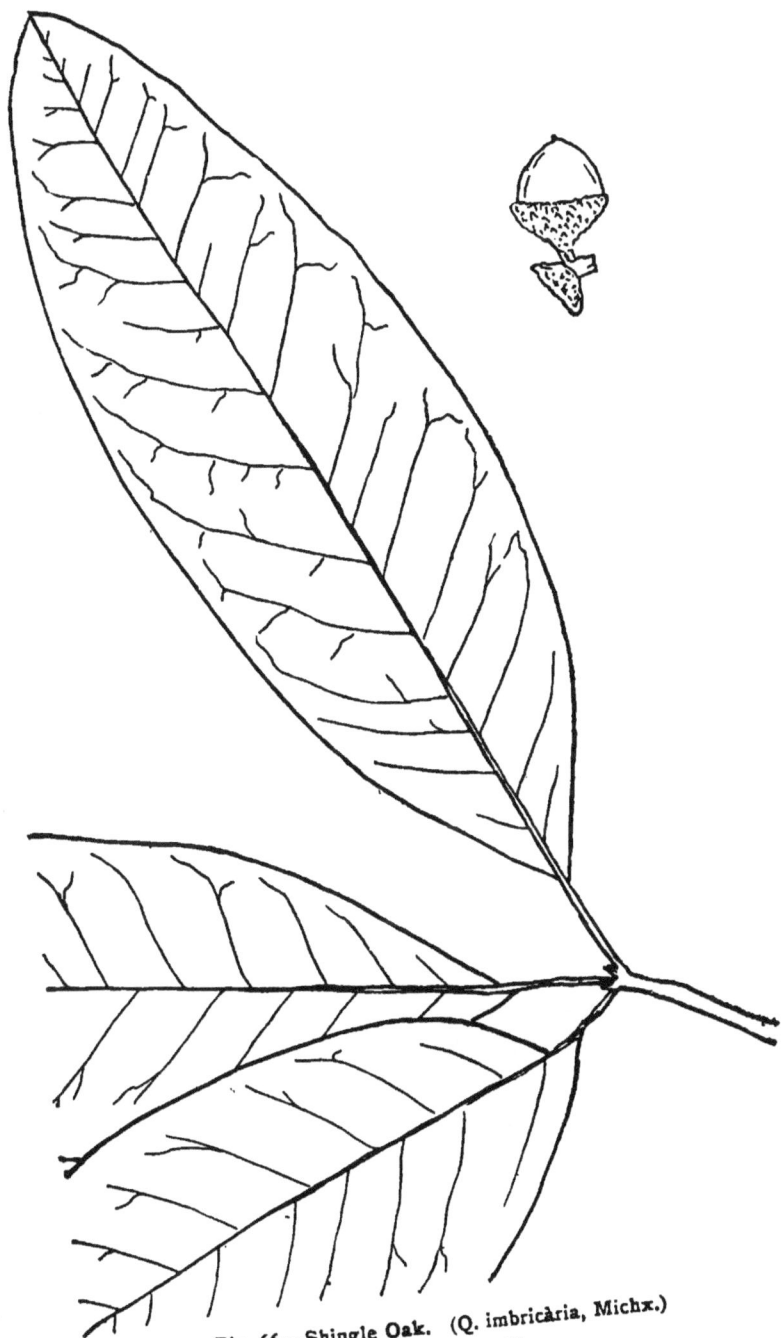

Fig. 66.—Shingle Oak. (Q. imbricària, Michx.)
NATURAL SIZE.

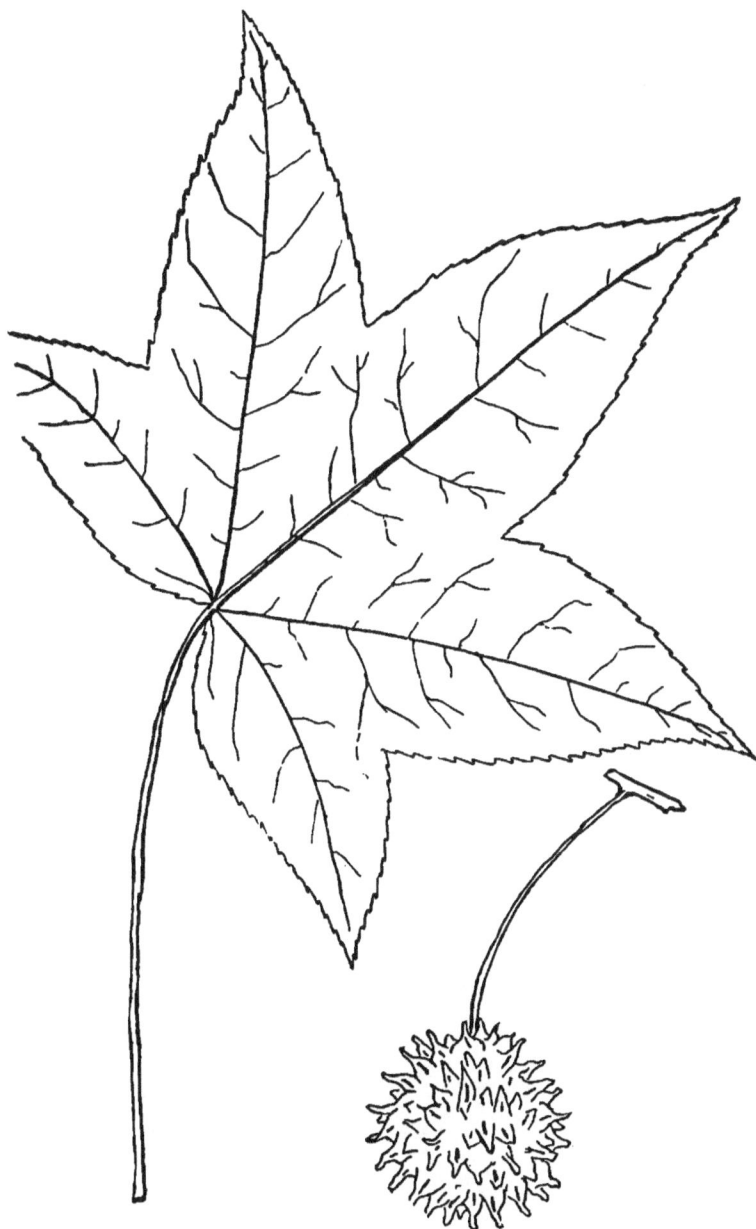

Fig. 67.—Sweet Gum. (L. styraciflua, L.)

NATURAL SIZE.

TREES WITH SIMPLE LEAVES

CONTINUED

LEAVES OPPOSITE

(EDGE ENTIRE)

B I

Fig. 68.—Flowering Dogwood. (C. flòrida, L.)
NATURAL SIZE.

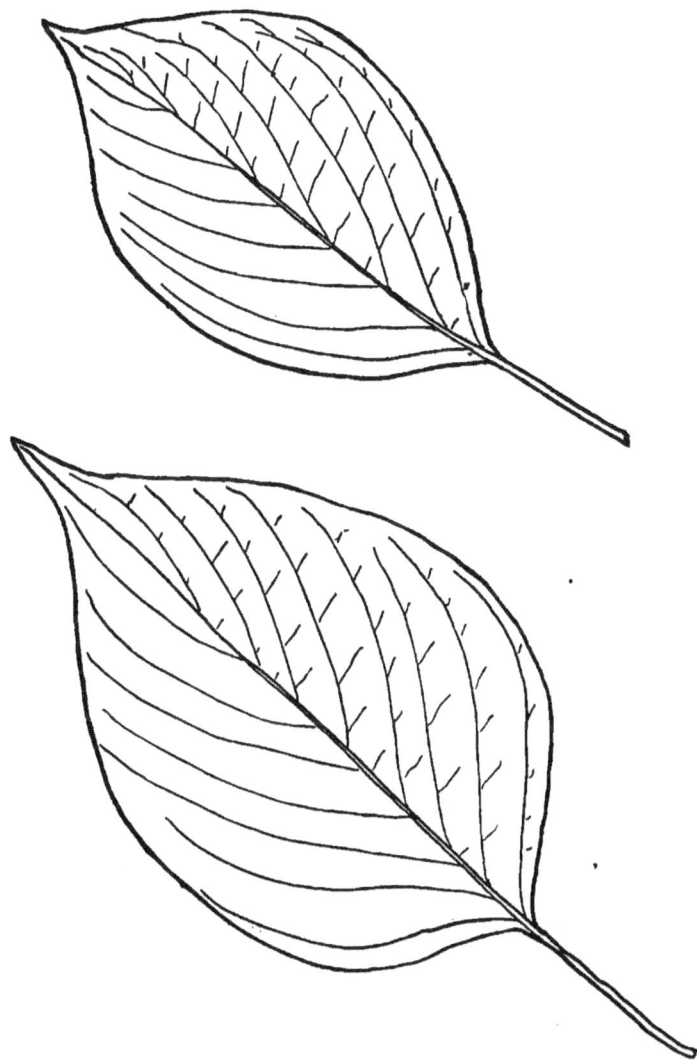

Fig. 69.—Alternate-leaved Dogwood. (C. alternifōlia, L. f.)
NATURAL SIZE.

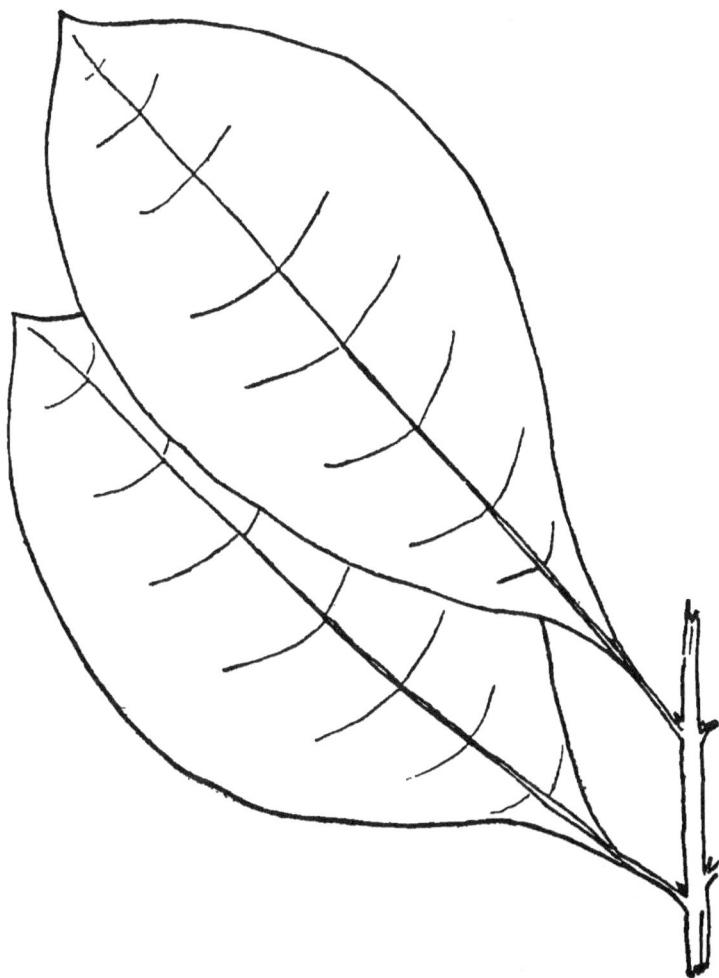

Fig. 70.—Fringe Tree. (C. Virginica, L.)
NATURAL SIZE.

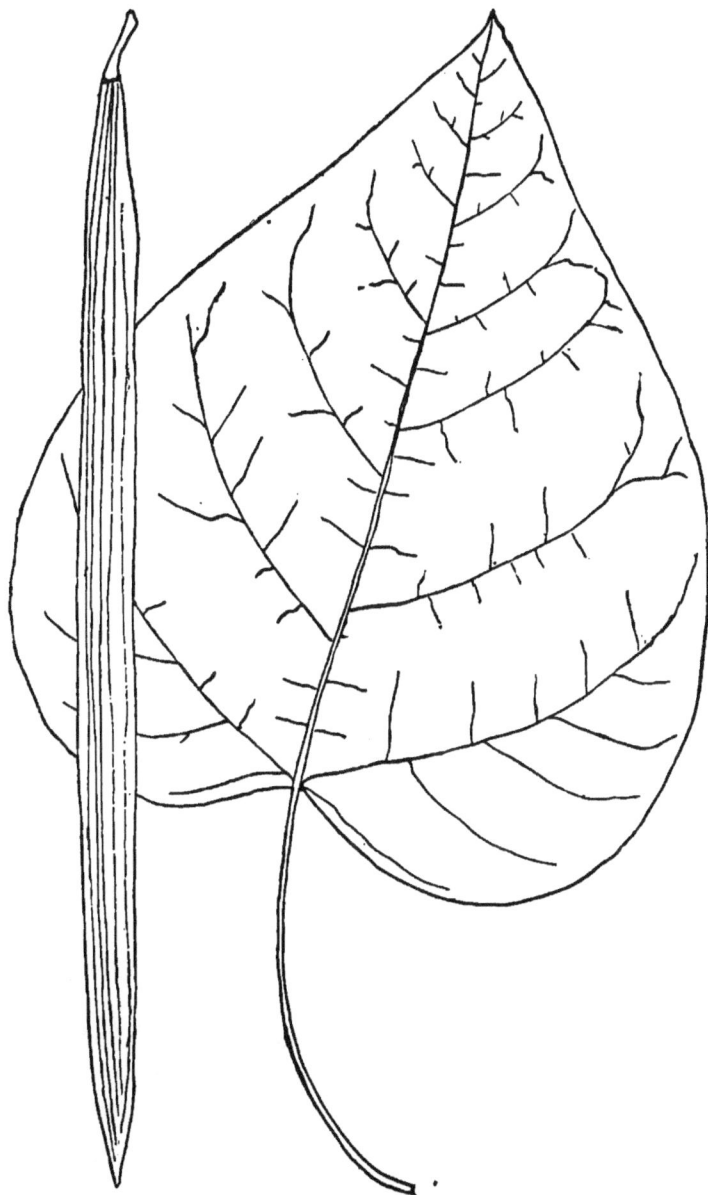

Fig. 71.—Catalpa. (C. bignonoides, Walt.)
LEAF AND FRUIT REDUCED ONE THIRD.

TREES WITH SIMPLE LEAVES

LEAVES OPPOSITE

CONTINUED

(EDGE TOOTHED)

B II

Fig. 72

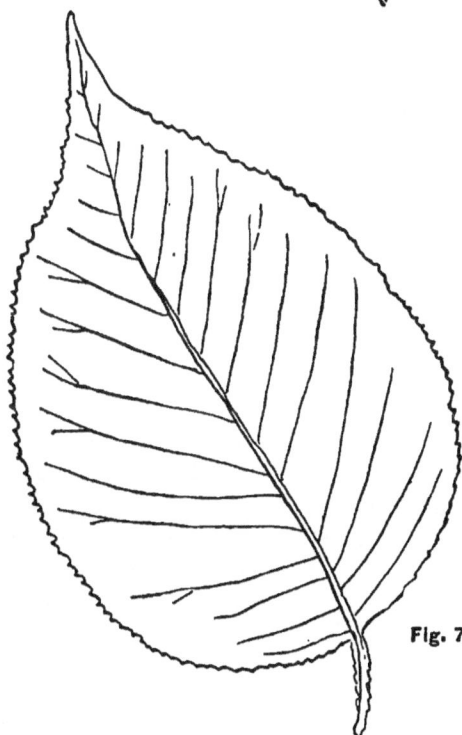

Fig. 73

Fig. 72.—Black Haw. (V. prunifòlium, L.)
Fig. 73.—Sweet Vibùrnum. (V. lentàgo, L.)
NATURAL SIZE.

TREES WITH SIMPLE LEAVES

LEAVES OPPOSITE

CONTINUED

(EDGE LOBED)

B III

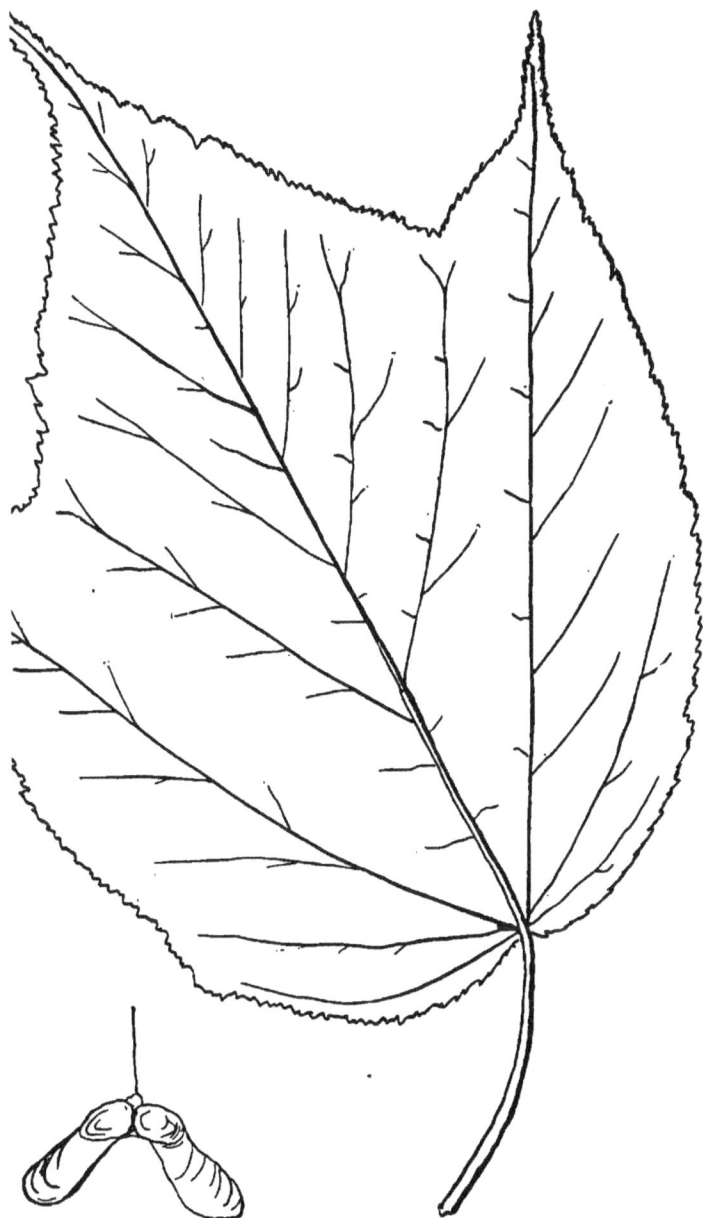

Fig. 74.—Striped Maple.   (A. Pennsylvànicum, L.)

NATURAL SIZE.

Fig. 75.—Sugar Maple. (A. sàccharum, Marsh.)
NATURAL SIZE.

Fig. 76.—Black Maple. (A. s., var. nigrum.)
NATURAL SIZE.

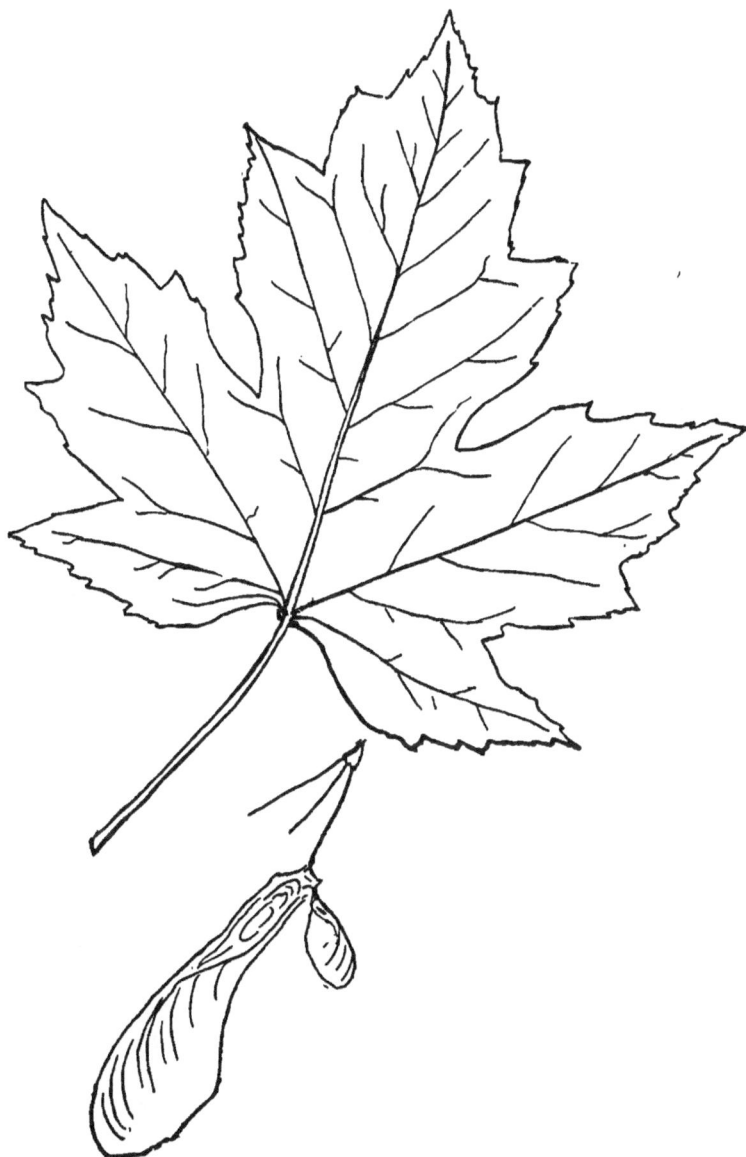

Fig. 77.—Silver-Leaf Maple. (A. saccharinum, L.)
NATURAL SIZE.

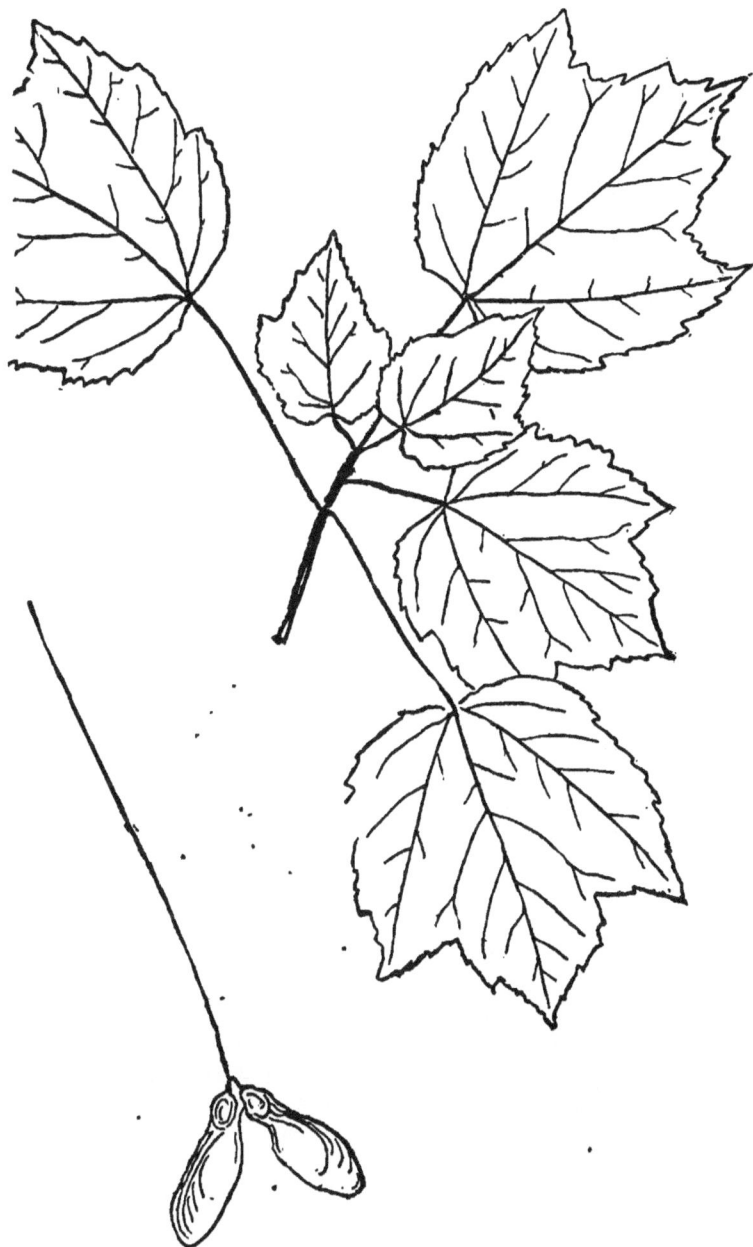

Fig. 78.—Red Maple.   (A. rubrum, L.)

NATURAL SIZE.

TREES WITH SIMPLE LEAVES

CONTINUED

LEAVES INDETERMINATE

C I

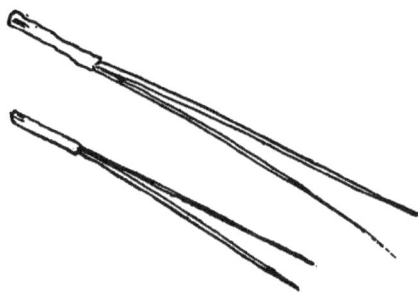

Fig. 80

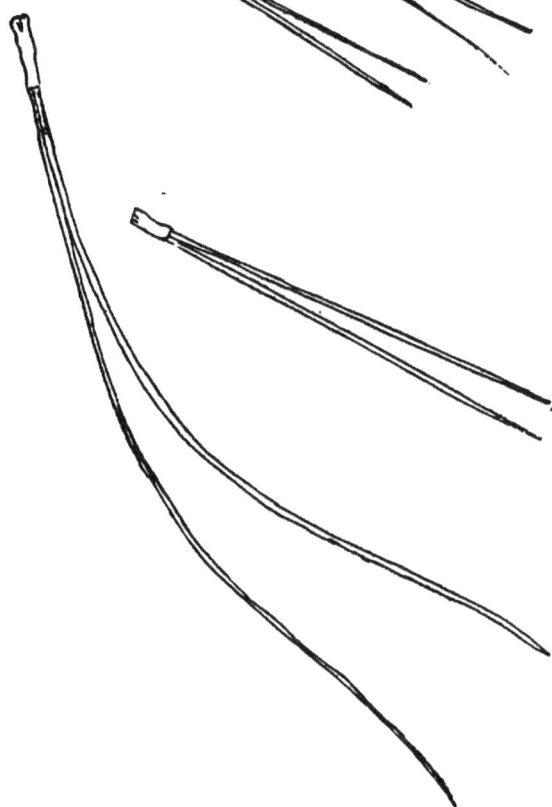

Fig. 81

Fig. 82

Fig. 79.—Gray Pine.   (P. Banksiàna, Lam.)
Fig. 80.—Scrub Pine.   (P. Virginiàna, Mill.)
Fig. 81.—Table Mountain Pine.   (P. pungens, Michx.)
Fig. 82.—Red Pine.   (P. resinòsa, Ait.)

NATURAL SIZE.

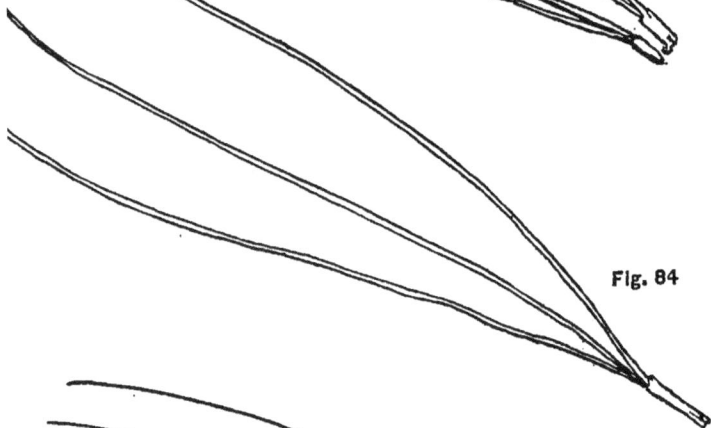

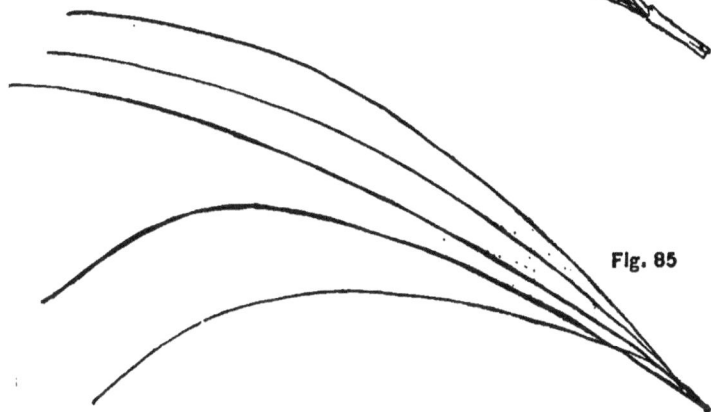

Fig. 83.—Yellow Pine.   (P. ecpinàta, Mill.)
Fig. 84.—Pitch Pine.   (P. rìgida, Mill.)
Fig. 85.—White Pine.   (P. Strobus, L.)

NATURAL SIZE.

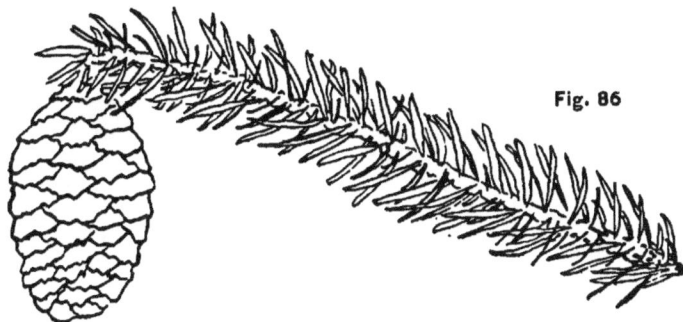

Fig. 86

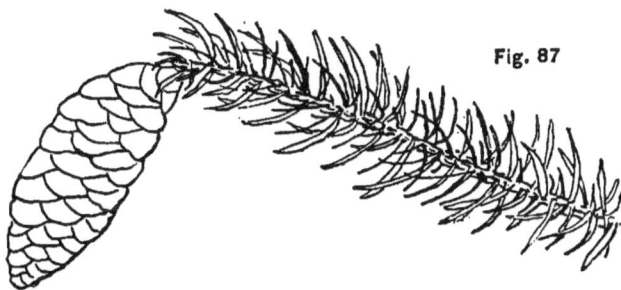

Fig. 87

Fig. 86.—Black Spruce.　P. Mariàna (Mill), B. S. P.
Fig. 87.—White Spruce.　P. Canadènsis (Mill), B. S. P.
NATURAL SIZE.

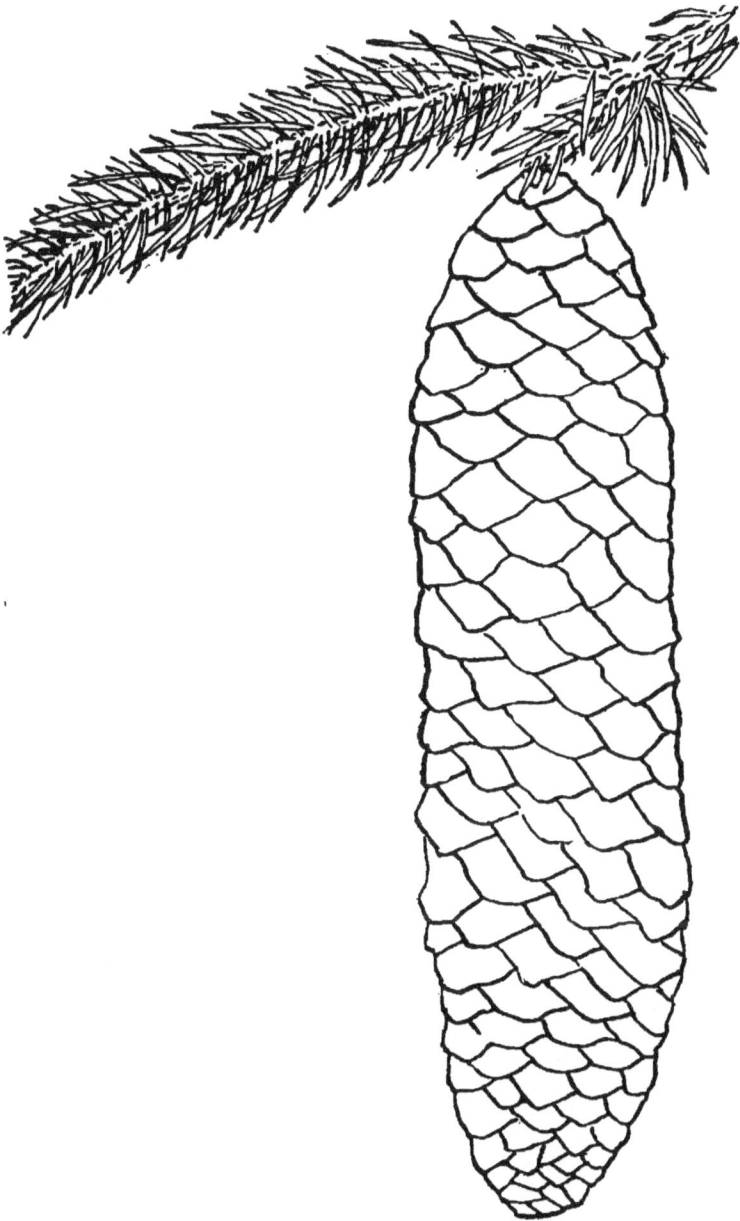

Fig. 88.—Norway Spruce. [P. excèlsa.]
NATURAL SIZE.

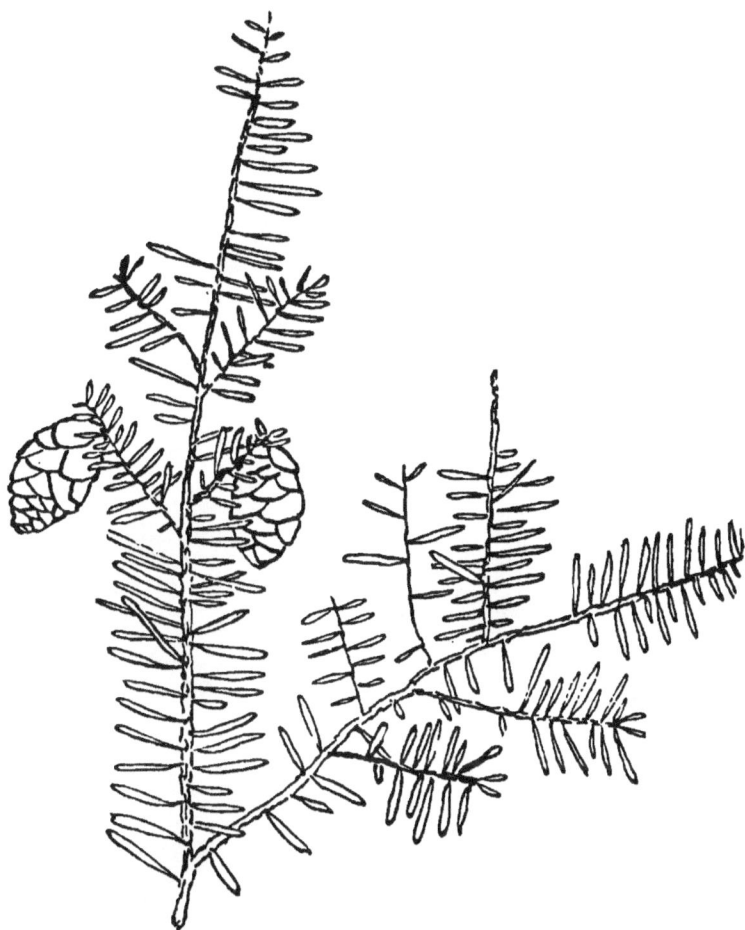

Fig. 89.—Hemlock. T. Canadènsis (L.), Carr.
NATURAL SIZE.

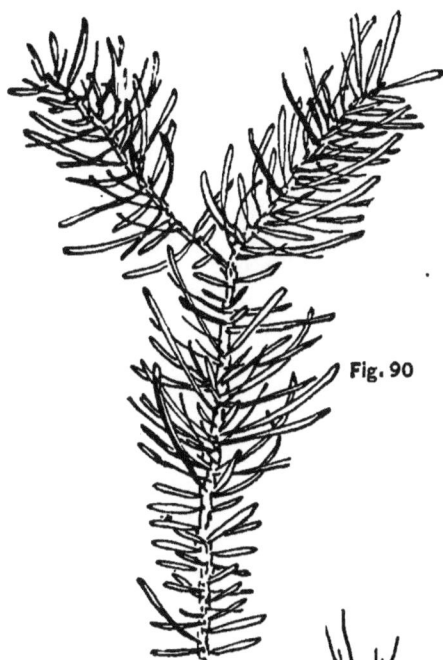

Fig. 90

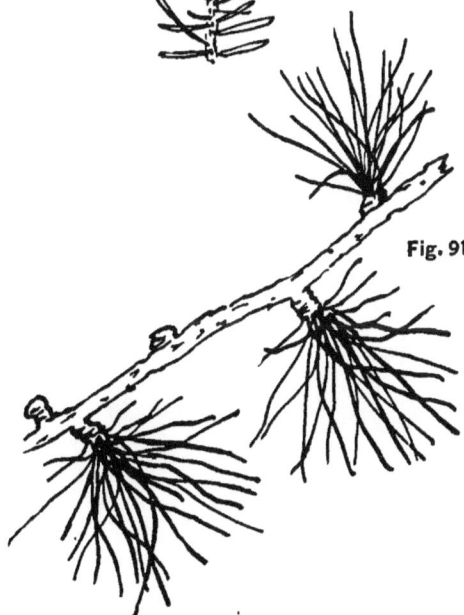

Fig. 91

Fig. 90.—Balsam Fir.   A. balsàmea (L.), Miller.
Fig. 91.—Larch.   L, laricina (Du Roi), Koch.
155   NATURAL SIZE.

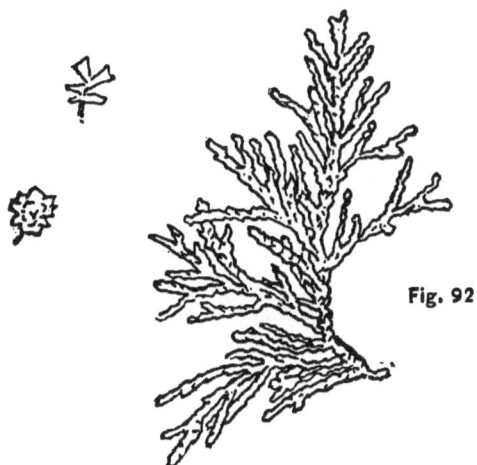

Fig. 92

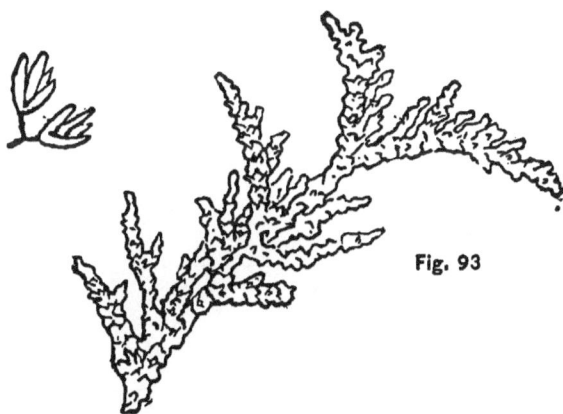

Fig. 93

Fig. 92.—White Cedar.   C. thyoìdes (L.), B. S. P.
Fig. 93.—Arbor Vitæ.   (T. occidentàlis, L.)
NATURAL SIZE.

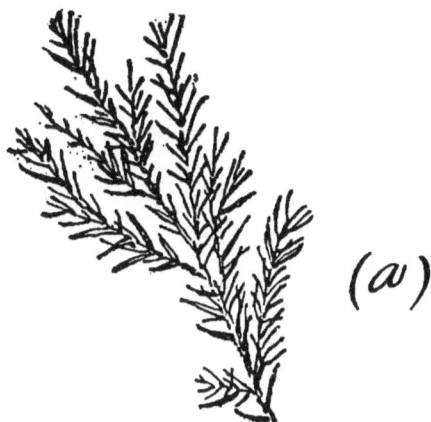

Fig. 94.—Red Cedar. (J. Virginiàna, L.)
a. Young. *b.* Old.
NATURAL SIZE.

REES WITH COMPOUND LEAVES
(FEATHER-SHAPED)

LEAVES ALTERNATE

(EDGE ENTIRE)

D I

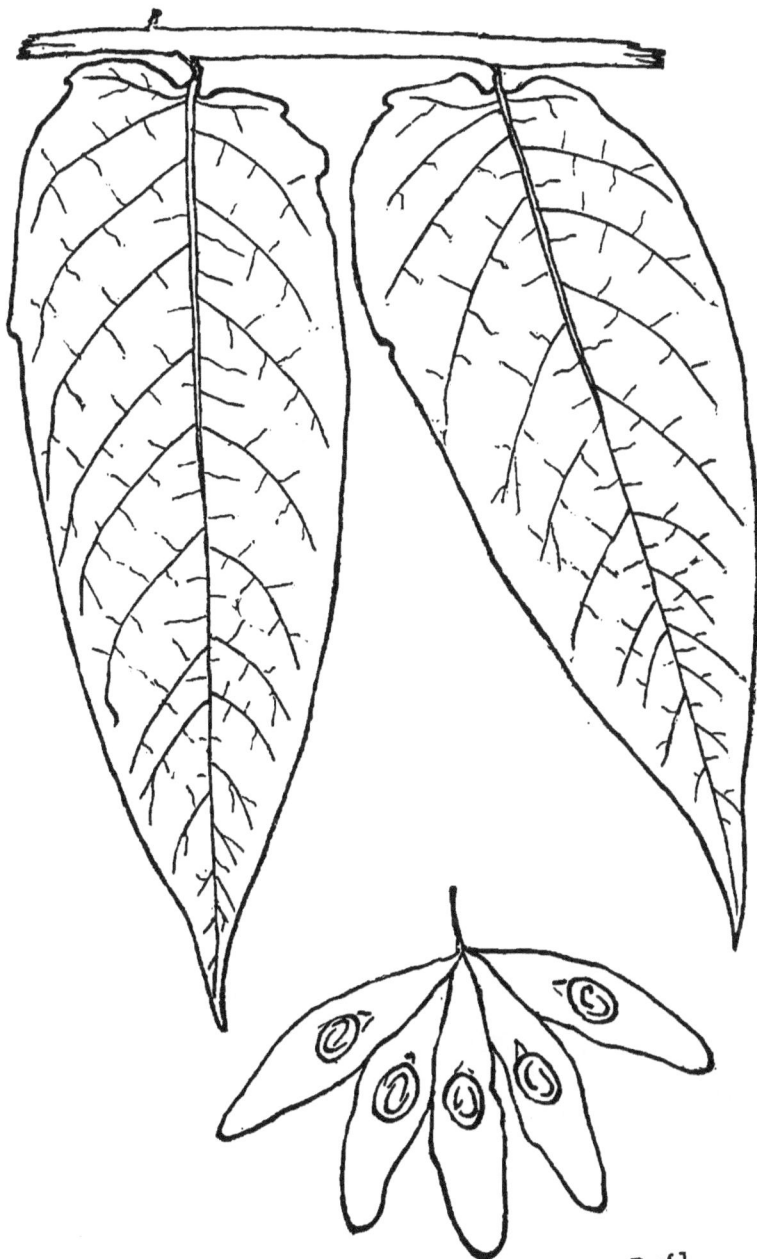

Fig. 95.—Ailànthus. [A. glandulòsa, Desf.]
NATURAL SIZE.

Fig. 96.—Locust.   (R. pseudacàcia, L.)
NATURAL SIZE.

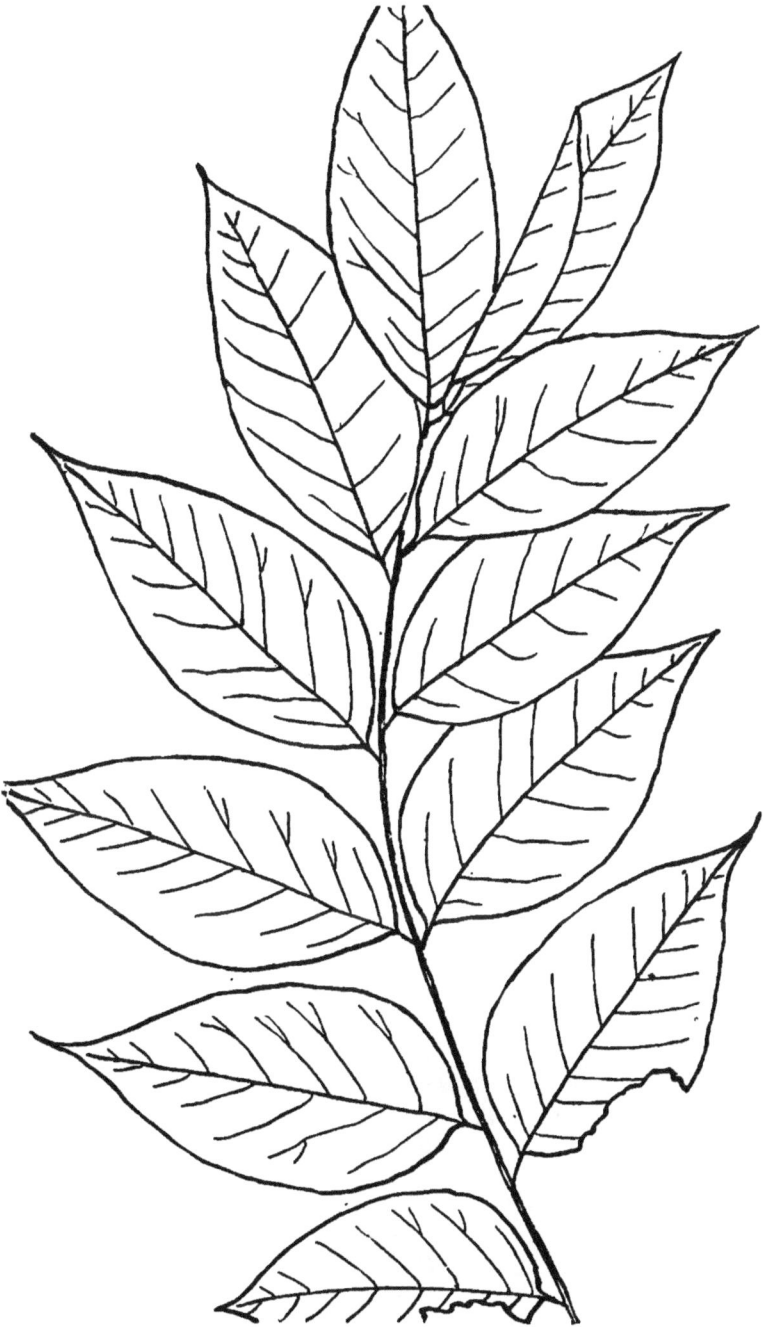

Fig. 97.—Kentucky Coffee Tree.   G. disicus (L.), Koch.

NATURAL SIZE.

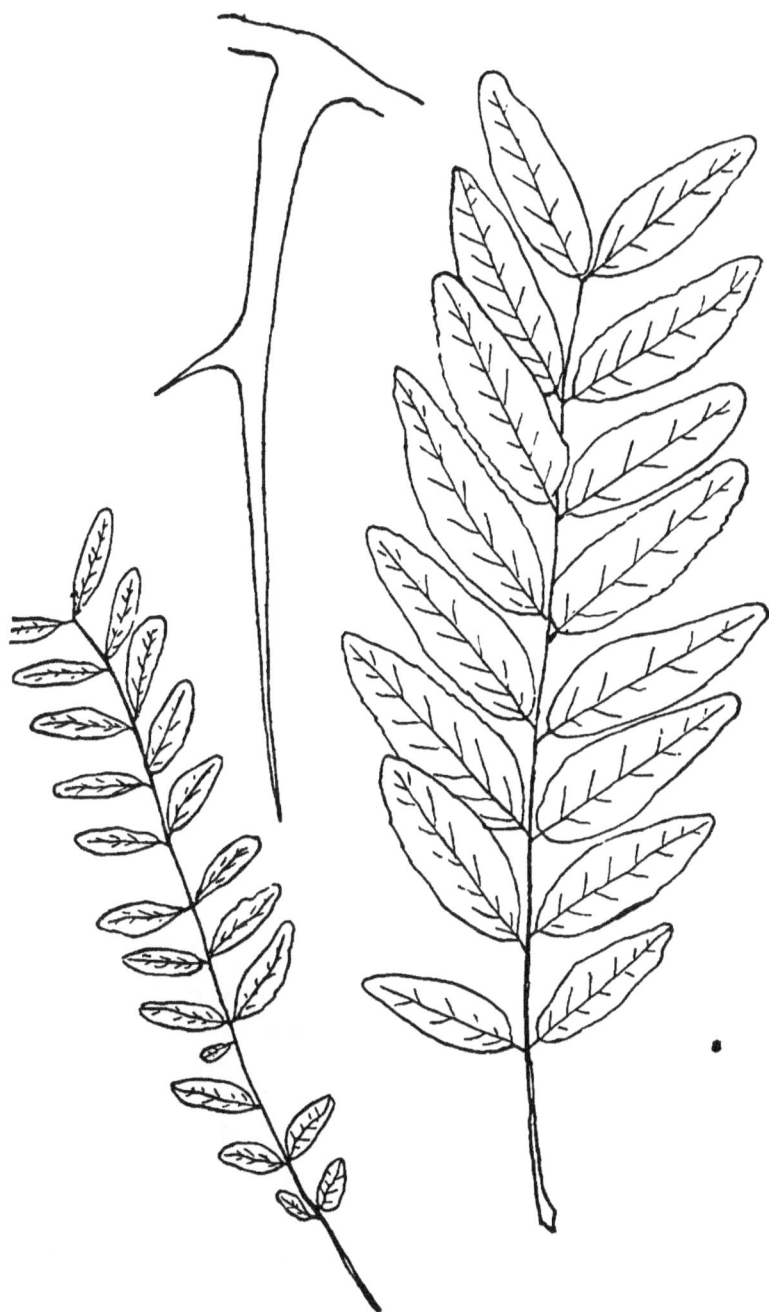

Fig. 98.—Honey Locust. (G. triacànthos, L.)

NATURAL SIZE.

TREES WITH COMPOUND LEAVES
(FEATHER-SHAPED)

LEAVES ALTERNATE

CONTINUED

(EDGE TOOTHED)

D II

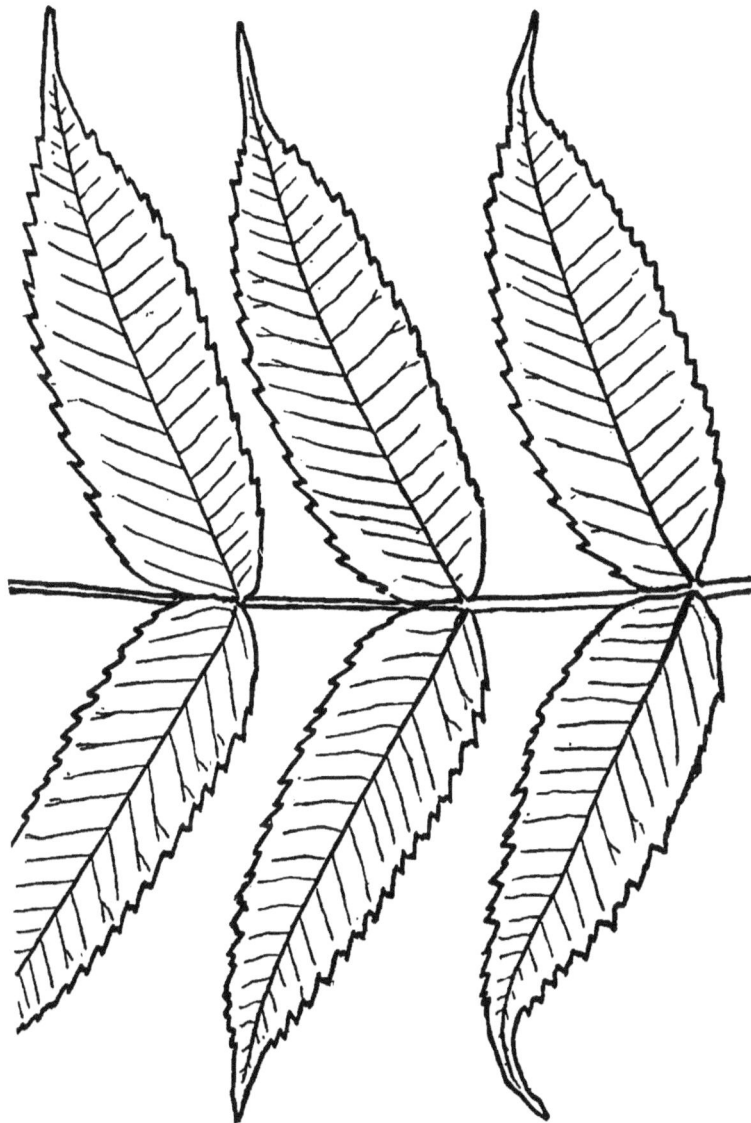

Fig. 99.—Stag-horn Sumach. (R. typhina, L.)
NATURAL SIZE.

.

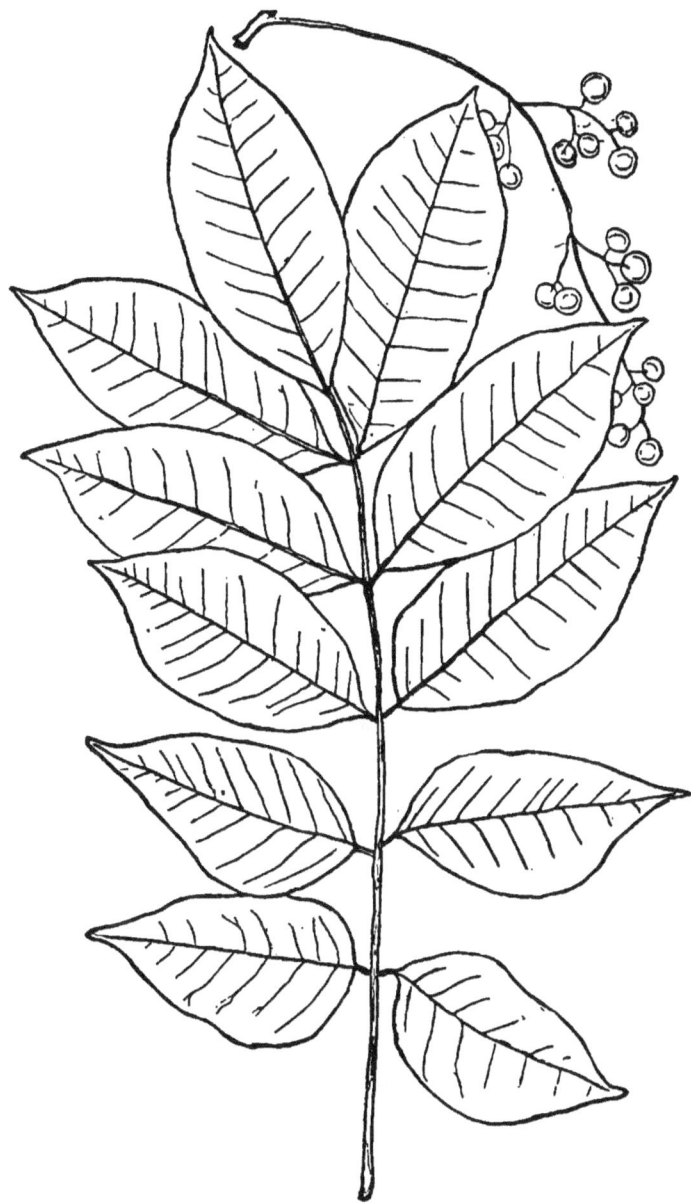

Fig. 100.—Poison Sumach. (R. venenàta, D. C.)
NATURAL SIZE.

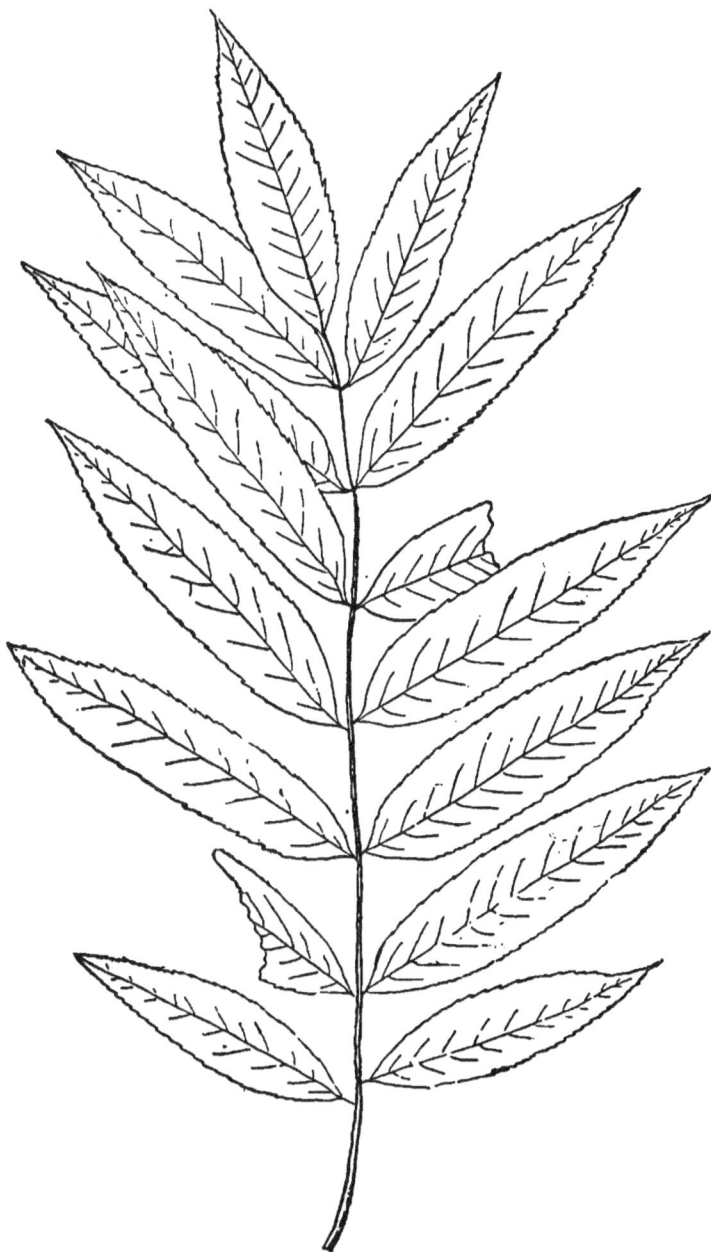

Fig. 101.—Mountain Ash.   (P. Americàna, D. C.)

REDUCED ONE FOURTH.

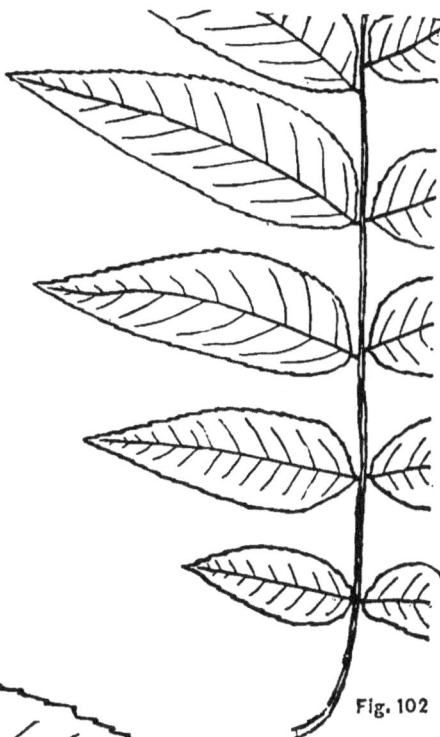

Fig. 102

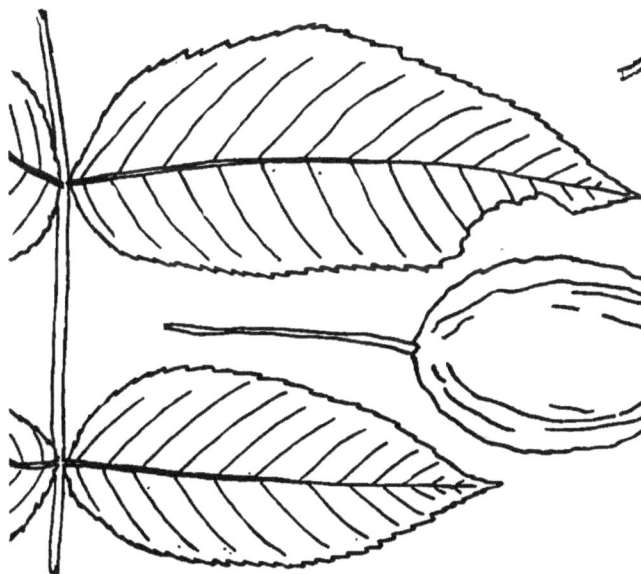

Fig. 103

Fig. 102.—Black Walnut. (J. nigra, L.)
Fig. 103.—Butternut. (J. cinèrea, L.)
179 LEAFLETS AND FRUIT REDUCED ONE THIRD.

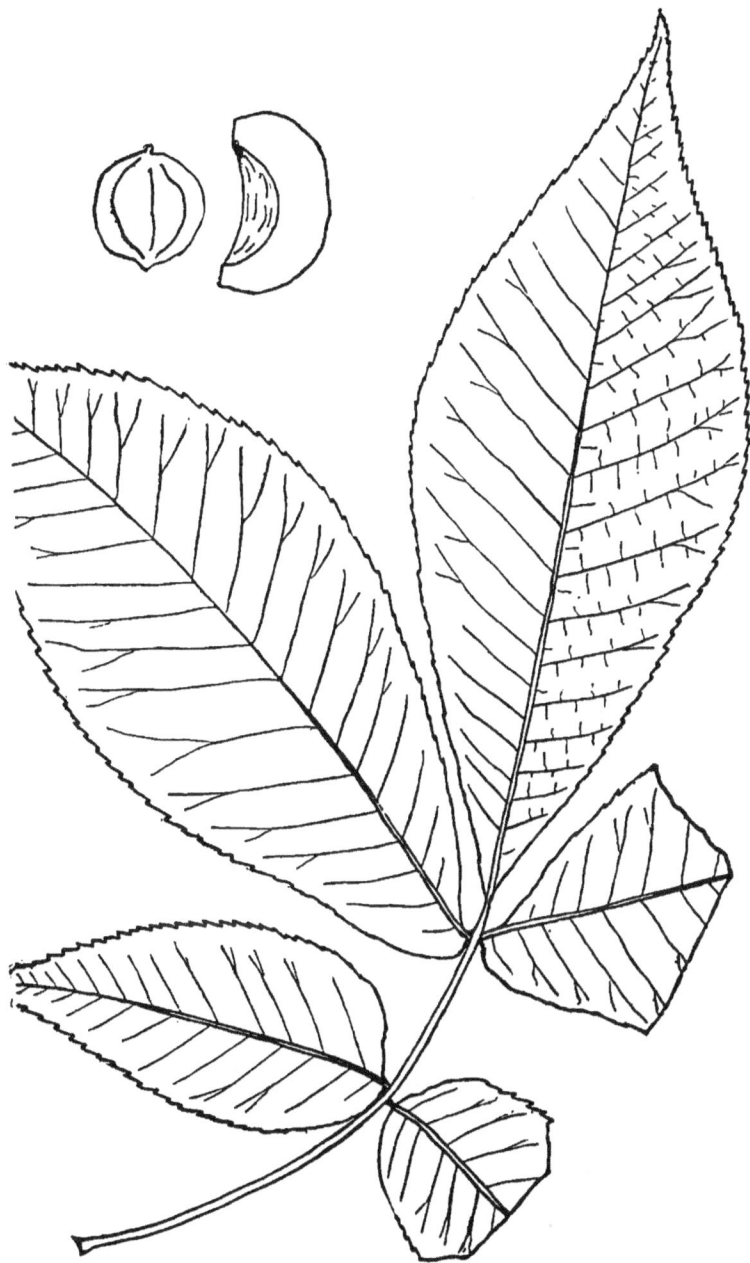

Fig. 104.—Shag-bark. H. ovàta (Mill), Britton.

LEAF AND FRUIT REDUCED ONE THIRD.

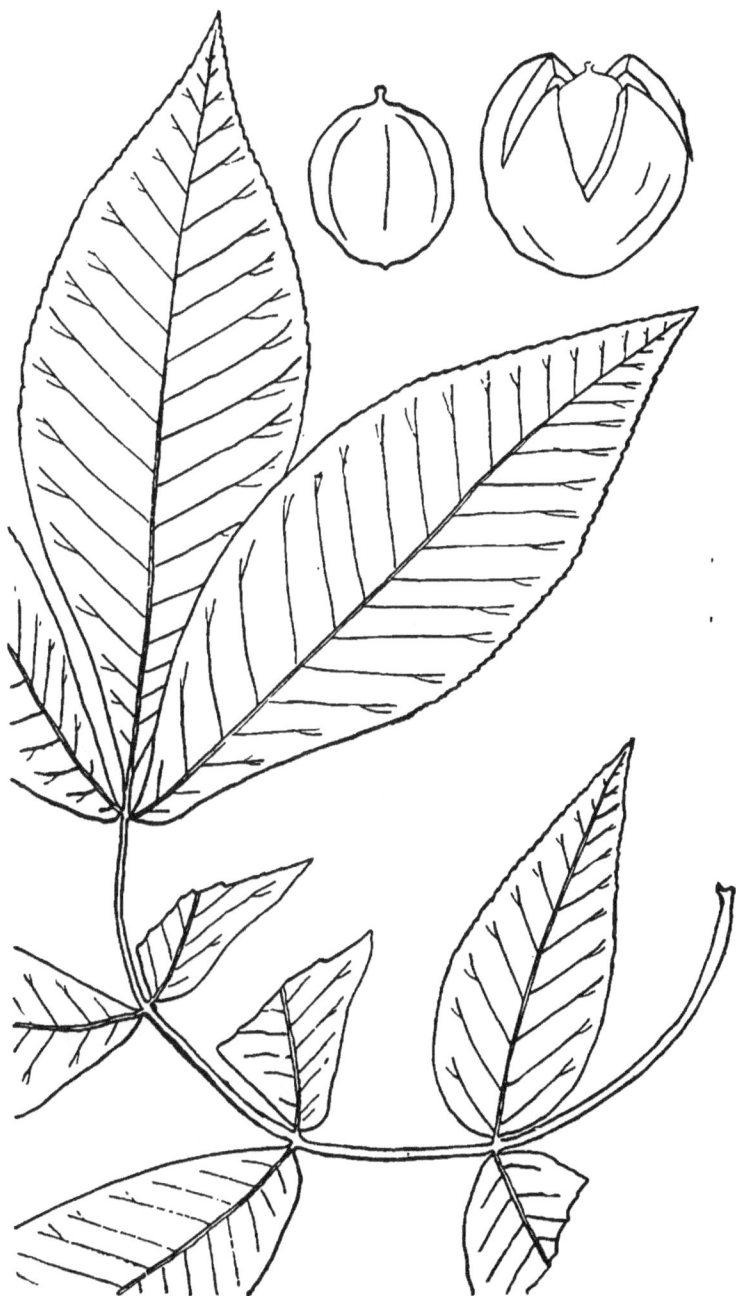

Fig. 105.—Mocker-nut. H. alba (L.), Britton.
LEAF AND FRUIT REDUCED ONE THIRD.

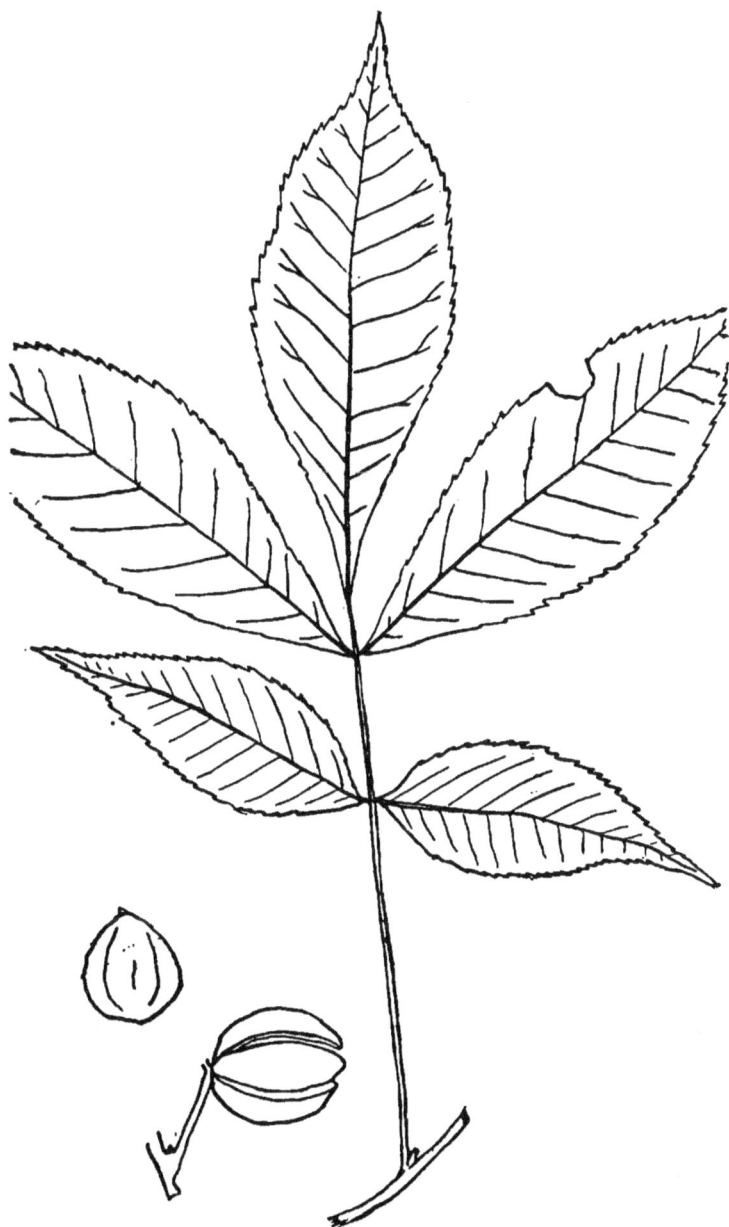

Fig. 106.—Small-fruited Hickory. H. microcàrpa (Nutt), Britton.
LEAF AND FRUIT REDUCED ONE THIRD.

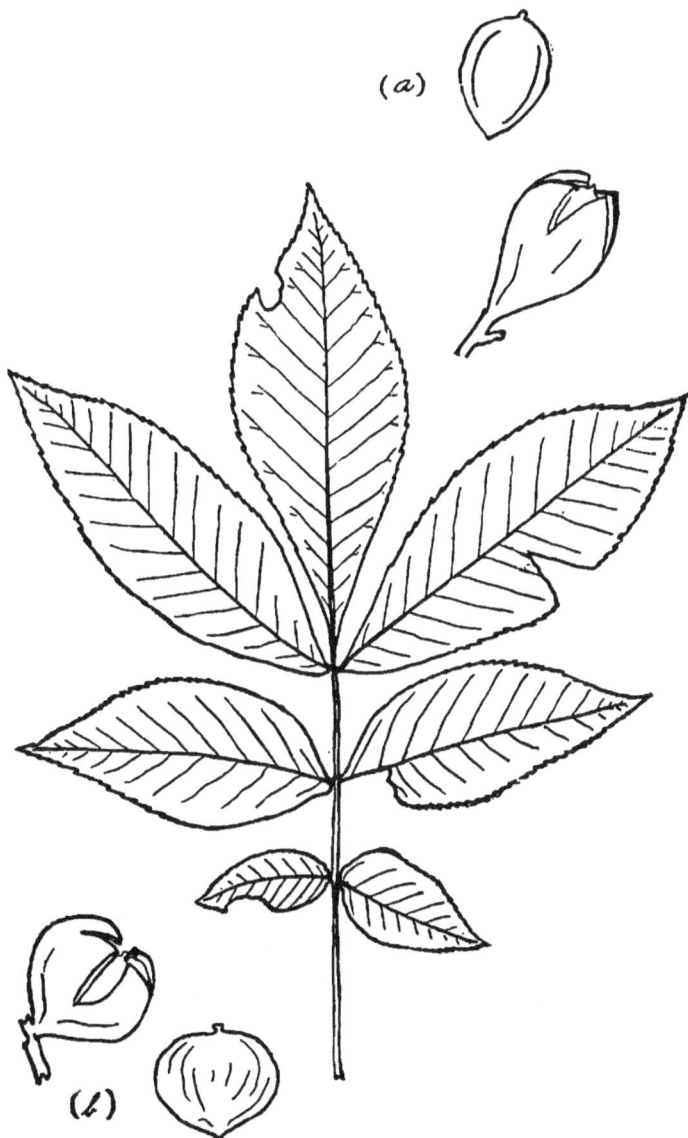

Fig. 107, *a* and *b*.—Pig-nut. H. glabra (Mill), Britton.
LEAF AND FRUIT REDUCED ONE THIRD.

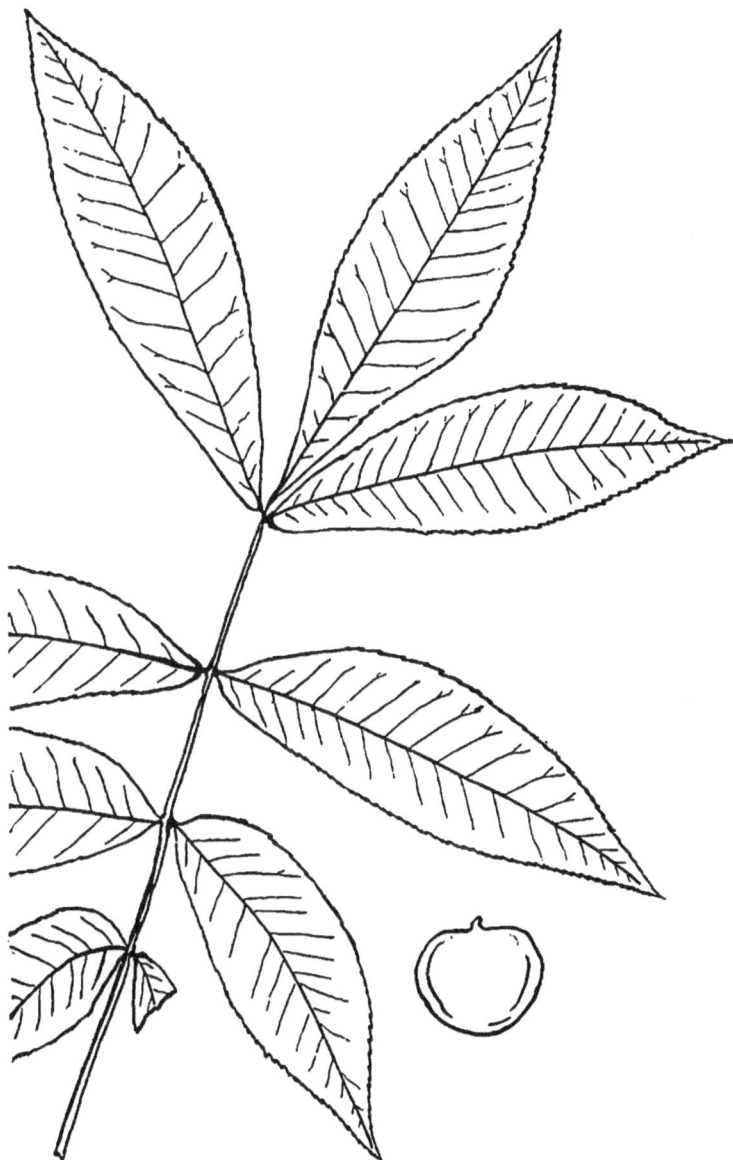

Fig. 108.—Bitter-nut. H. minima (Marsh), Britton.
LEAF AND FRUIT REDUCED ONE THIRD.

TREES WITH COMPOUND LEAVES

(FEATHER-SHAPED)

CONTINUED

LEAVES OPPOSITE

(EDGE ENTIRE OR TOOTHED)

E I, II

Fig. 109.—Ash-leaved Maple. (N. aceroïdes, M.)
NATURAL SIZE.

Fig. 110.—White Ash. (F. Americàna, L.)
LEAF AND FRUIT REDUCED ONE THIRD.

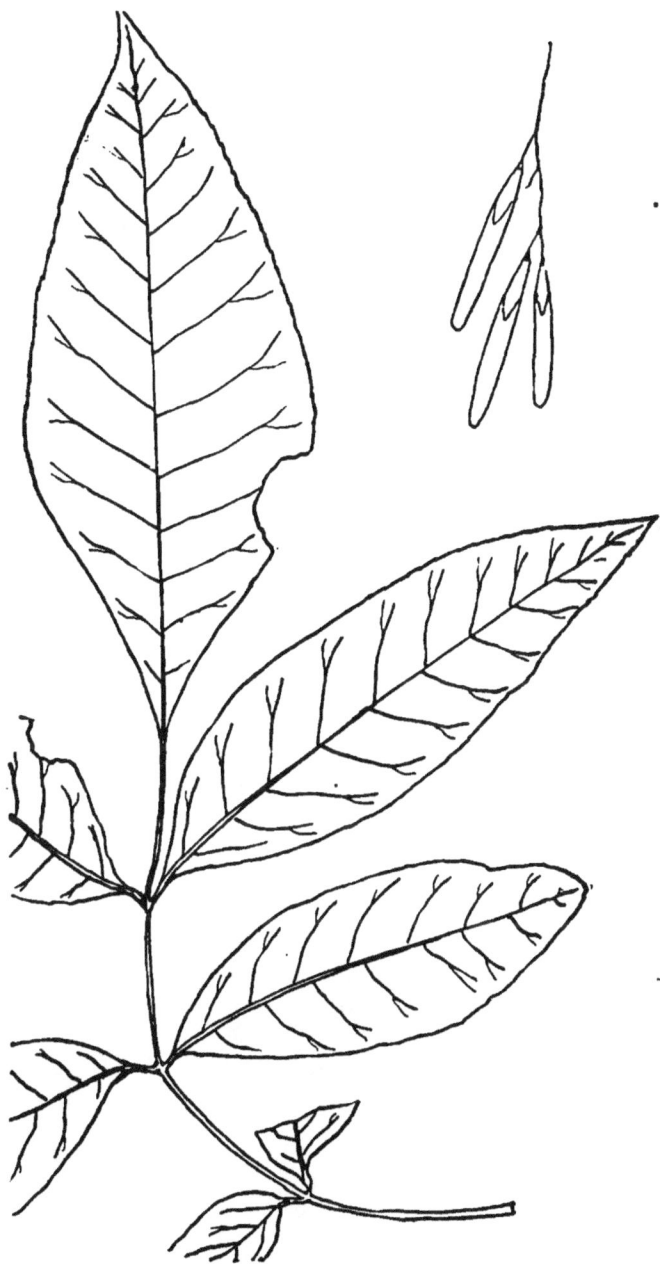

Fig. 111.—Red Ash. (F. pubèscens, Lam.)

LEAF AND FRUIT REDUCED ONE THIRD.

Fig. 112.—Green Ash. (F. viridis, Michx., f.)
LEAF AND FRUIT REDUCED ONE THIRD.

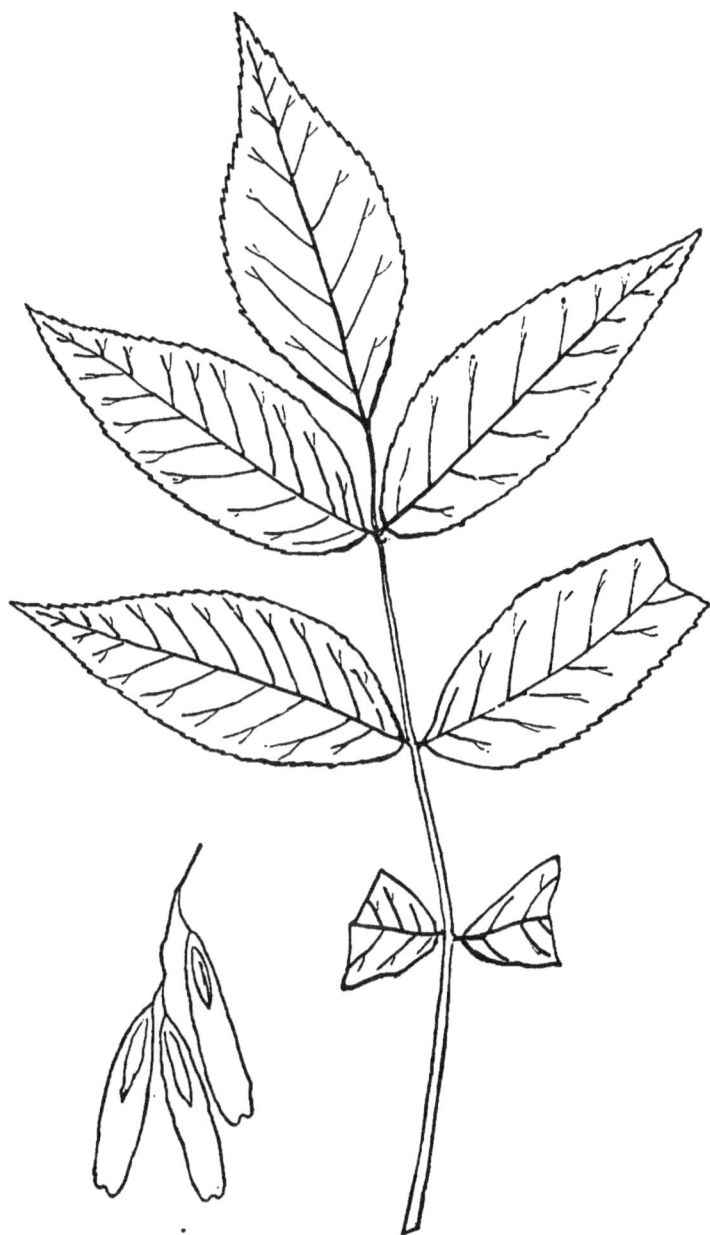

Fig. 113.—Blue Ash. (F. quadrangulàta, Michx.)
LEAF AND FRUIT REDUCED ONE THIRD.

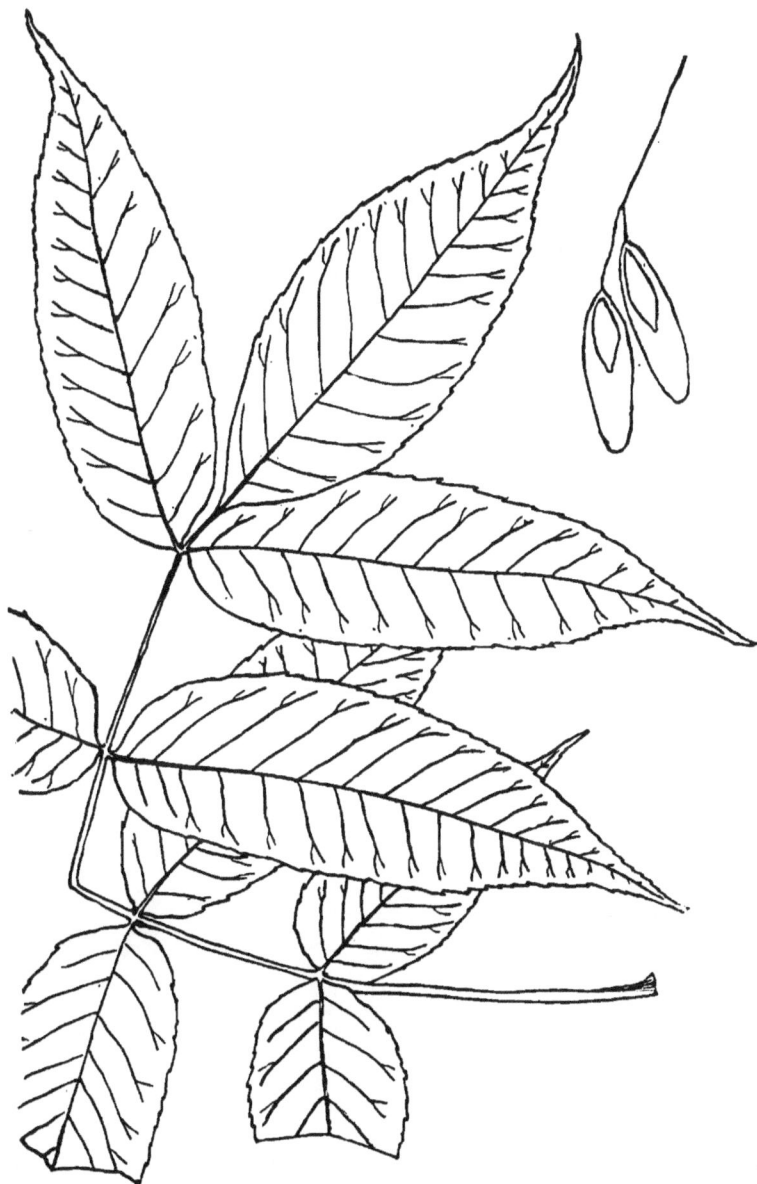

Fig. 114.—Black Ash. (F. sambucifòlia, Lam.)
LEAF AND FRUIT REDUCED ONE THIRD.

TREES WITH COMPOUND LEAVES

(HAND-SHAPED)

LEAVES OPPOSITE

(EDGE TOOTHED)

F I

Fig. 115.—Sweet Buckeye. (Æ. flava, Ait.)
REDUCED ONE THIRD.

Fig. 116.—Ohio Buckeye. (Æ. glabra, Willd.)
REDUCED ONE THIRD.

NAMES OF OMITTED AND COMPARATIVELY UNIMPORTANT TREES.

Tilia heterophýlla. Vent.
 pubèscens. Ait.
Pyrus angustifòlia. Ait.
Cratægus coccìnea, L. var. mollis.
 crus-galli, L. pyracanthifòlia.
Amelánchier Canadènsis, L. var. T. and
 G.
Ulmus racemòsa. Thomas.
Celtis occidentàlis, L. var. crassifòlia.
Salix amygdaloìdes. Anders.
Quercus macrocàrpa, Michx. var. olivæfòrmis.

Quercus coccìnea, Wang. var. ambìgua,
 Gray.
(Of nine hybrid oaks, most are outside our
 limits or entirely local.)
Catàlpa speciòsa. Ward.
Roblnia viscòsa. Vent.
Gledìtschia triacànthos, L. var. inèrmis,
 and var. brachycàrpos.
Rhus týphina, L. var. laciniàta.
Pyrus sambucifòlia
Hicòria sulcàta (Willd). Britton.

INDEX OF TREES.

The names of genera are given in SMALL CAPITALS, of species and varieties in Roman type. The names of introduced species are enclosed by brackets.